Quality and Power in the Supply Chain: What Industry Does for the Sake of Quality

Quality and Power in the Supply Chain:

What Industry Does for the Sake of Quality

James Lamprecht, Ph.D.

Boston   Oxford   Auckland
Johannesburg   Melbourne   New Delhi

Recognizing the importance of preserving what has been written, Butterworth–Heinemann prints its books on acid-free paper whenever possible.

 Butterworth–Heinemann supports the efforts of American Forests and the Global ReLeaf program in its campaign for the betterment of trees, forests, and our environment.

Library of Congress Cataloging-in-Publication Data
Lamprecht, James L., 1947–
    Quality and power in the supply chain : what industry does for the sake of quality / James Lamprecht.
        p.   cm
    Includes bibliographical references and index.
    ISBN 0-7506-7343-5 (alk. paper)
    1. Industrial procurement.   2. Quality assurance.   3. Business networks.   I. Title.
HD39.5 .L36 2000
658.4′012—dc21   00-031226

British Library Cataloguing-in-Publication Data
A catalogue record for this book is available from the British Library.

The publisher offers special discounts on bulk orders of this book.
For information, please contact:

Manager of Special Sales
Butterworth–Heinemann
225 Wildwood Avenue
Woburn, MA 01801-2041
Tel: 781-904-2500
Fax: 781-904-2620

For information on all Butterworth–Heinemann publications available, contact our World Wide Web home page at: http://www.bh.com

10 9 8 7 6 5 4 3 2 1

Printed in the United States of America

Contents

Part V Consequences of Standardization

14. On the Origin of Procedures   171

15. Writing Procedures   179

Part VI Conclusion

16. By Way of Conclusion: Dos and Don'ts   201

List of Tables

Preface

When Safeway, a grocery store operating mostly in the Southwest, initiated its "service-with-a-smile" policy, some workers said that management had gone too far with its "superior customer service" program. To ensure that employees were abiding by the new policy, Safeway regularly used undercover shoppers to evaluate each worker's performance. "Negative evaluation can lead to remedial training, disciplinary letters and termination."[1] The program, which requires all employees to smile, greet, and make eye contact with every customer who walks in the store, has led to some unexpected problems for some employees who say they are propositioned daily by men who mistake their "company-mandated friendliness as flirtatiousness." Naturally, Safeway officials deny that their employees are forced to "smile in a certain way," and suggest that the unfortunate and unusual experience can perhaps be alleviated by offering more training to deal with unruly customers. Still, this is but one example of a corporate quality policy that has gone awry. Many others could be cited.

This book was born out of an attempt to reconcile what I perceive to be two divergent, perhaps even contradictory, worlds. On the one hand is the world of quality and managerial fads; a world promoted by quality professionals and the quality industry, with its seminars, conferences, software packages, books, magazines, certification programs, and "approved" specialized courses. This world also consists of a multitude of arbitrary and, at times, irrational demands imposed on thousands of businesses and suppliers by an ever increasing number of international standards, state and federal regulations, and powerful corporations. The antithesis to this world of mission statements, quality policies, leadership training, six-sigma statistical techniques, procedure writing, and certification programs is a separate reality that has been captured by the cartoonist Scott Adams with his immensely popular character known as Dilbert. Wherever I visit organizations, I have come across cartoon strips of *Dilbert* taped on office walls or doors. No topic is too sacrosanct for Scott Adams' cast of characters; they poke fun at mission statements, quality meetings, leadership training, ISO 9000 certification, empowerment, marketing, design engineers, the boss, corporate policy, and more. The *Dilbert* phenomenon is not only national but also international. Everyone can relate to the corporate (in)sanity described by Scott Adams' cast of characters. By poking fun at corporate America, Scott Adams allows millions of employees to temporarily relieve the stress caused by the everyday irrationality,

---

[1] "Some Safeway Clerks Object to Service-with-Smile Policy," *The Press-Enterprise,* September 2, 1998, p. A3.

nonsense, and occasional absurdity of corporate leaders who, in their attempt to satisfy yet another customer requirement, send their staff searching for the latest "quick fix."

Quality and Power in the Supply Chain attempts to bridge the gap between the vast literature of quality fads including the recent fad of ISO 9000 international standards and Scott Adams' humorous description of these worlds.[2] Unlike Adams, who simply pokes fun at the irrationality and at times idiotic behavior of corporate policies, I wanted to trace the origin of some of today's managerial and quality ideologies, show how quality management movements evolve, explain how the quality industry and its profession popularize, promote, and benefit from fads, and finally suggest that despite all the efforts and propaganda published by the quality industry, there is a separate reality to "quality" which clearly demonstrates that managerial principles of quality are but one small limiting factor to corporate success.

This is a book about many subjects rarely if ever covered by management or quality professionals and practiced by thousands on a daily basis for the sake of quality. The book covers many topics such as the abuse of quality as practiced by powerful organizations on their suppliers; the limits of quality in an age of corporate takeovers and downsizing; the relativity of quality across cultures; the business of quality; the quality profession and its role in promoting various fads including international standards, bureaucratic registrars, and standardization (ISO 9000); the role of power in customer–supplier relations, the economic survival of companies; and plain stupidity, including the occasional stupidity of customers, of burdensome procedures, and of quality fads in general. This eclectic, yet related, series of topics is the foundation of the book. The broad nature of the subject matter offered me little alternative but to gather the varied subjects into six interrelated parts.

Part I, Prologue, consists of Chapters 1 and 2. In it, the general premise of abuse of power for the supposed sake of quality and the absurdity that occasionally ensues from the forced implementation of arbitrary rules and regulations is introduced. Chapters 1 and 2 analyze the various abuses exercised by corporations on their suppliers, employees, and customers, supposedly for the sake of quality improvement. Chapter 1 describes how power is used by certain organizations to interfere or otherwise control the manufacturing process of their suppliers. Chapter 2 illustrates how power and bureaucracy allow some certification agencies, known as registrars, to go beyond the intent of the ISO 9000 standards and occasionally impose unreasonable demands on organizations that want to achieve registration.

Part II, The Limits of Quality: Essays on a Separate Reality, contains Chapters 3 through 7. These five chapters examine a series of issues rarely if ever covered by quality professionals. Chapter 3 explores some of the difficulties presented by the relativity of quality. Chapter 4 demonstrates that the success of a company, measured by how long it has been in business, has, more often than not, little to do with its mission statement, corporate philosophy, or quality policy. Chapter 5

---

[2]Scott Adams' most popular book is *The Dilbert Principle*, HarperBusiness, New York, 1996.

expands on the theme presented in Chapter 4 and shows that quality is but one element of success. Chapter 6 questions what is meant by "the customer" and asks whether it is wise to serve all customers equally. Chapter 7 reviews many cases of what can euphemistically be called absurdity as it relates to quality (in)efficiency.

Part III, Colbertism and the Dawn of Power in Customer–Supplier Relations, traces some of the historical antecedents to the use and abuse of power in customer–supplier relations. In Chapters 8, 9, and 10, I demonstrate how principles of manufacturing standardization had already been developed more than 330 years ago to address the demands of one powerful customer, Colbert. Colbert's attempt at establishing French manufacturing standards is interesting because it not only reveals how similar Colbert's approach was to the current ISO 9000 phenomenon, but it also demonstrates that early standards developers may have been less dogmatic and more farsighted than today's members of international standardization committees. Chapter 9 traces the origin of process standardization and standardization in general to the "quest for repeatability." The role the military had in imposing this quest for manufacturing repeatability and its impact on the quality profession and organizations in general are presented in Chapter 10.

Chapters 11, 12, and 13 make up Part IV, The Age of Standardization, which describes the origins of the standardization movement. Chapter 11 presents the pros and cons of standardization. Chapter 12 reviews the ISO 9000 phenomenon and explains how military standards were transformed into the ISO 9000 series for consumption by the private sector. Chapter 13 reviews how the quality profession and the quality industry in general helped promote the lucrative business that became known as the ISO 9000 phenomenon.

Part V, Consequences of Standardization, consists of Chapters 14 and 15. Because one of the by-products of the ISO 9000 phenomenon has been the generation of documented procedures, Part V analyzes the world of procedure writing. Chapter 14 offers a brief history of the origin of procedures and offers suggestions on how to integrate "working knowledge." Chapter 15 explores the world of procedures: what they are, when they are needed, how to write them, and when they should be avoided.

Part VI concludes with Chapter 16, which offers a brief look at the future of the quality profession and the businesses it represents. Suggestions are offered on what to do and what to avoid doing in this age of fads, quick solutions, and occasional absurd behavior.

I wish to express my thanks to Michael Forster of Butterworth-Heinemann for accepting this project, as well as the editors and reviewers, particularly Raynard Van Der Westhuizen of South Africa, for their time, effort, encouragement, and patience. I also wish to thank my wife Shirley for reading yet another manuscript and offering some thoughtful suggestions.

James Lamprecht
Temecula, California
January 2000

Part I

Prologue: On Power, and Its Impact on Customer–Supplier Relations

CHAPTER 1

Power and Its Impact on Customer–Supplier Relations

If one of your goals in life is to avoid unnecessary headache and heartache, then avoid QS 9000 like the plague.

—*Mike*

If one of your goals in life is to be happy, avoid automotive all together.

—*Marc*

These words of wisdom were found on a private web page and offered to an Italian engineer named Paolo who was inquiring about whether or not his company should obtain QS 9000 certification—an extensive and costly quality management system required by the automotive industry from all of their suppliers.

Introduction

As a young teenager raised until 1964 in Casablanca, Morocco (a former French colony), I always dreamed of having an electric guitar. Influenced by the success of such groups as The Ventures, The Shadows (the British equivalent of The Ventures), and, later, The Beatles, I thought that I too could one day perform on stage. Fender guitars were my favorite, but since I could not afford even the cheapest model, I would try to console myself by admiring the many exorbitantly priced models displayed at a well-known music store. One day, at age 14, I decided that I had waited long enough; I had to have an electric bass. I went to see a craftsman who own a small furniture shop. What was unique about this cabinetmaker—by today's American standards—was that he would custom design furniture (tables, couches, chairs, cabinets, complete living rooms, etc.). Customers would come to his shop and place an order for a Louis XV dresser or other period furniture. The owner would draw a few sketches, generally on the back of an envelope; the customer would look at the sketches, offer a few suggestions and comments regarding the shape or dimensions; a new diagram or sketch would be drawn; and, within a few minutes, the customer would say, "Yes,

3

that is what I want. How much would it cost?" Of course, purchasing custom-designed furniture is generally considered unthinkable nowadays simply because the cost would be considered, by most, prohibitive.[1]

Having decided that I was going to have my electric guitar after all, I grabbed an album of my favorite group (which had a picture of my dream guitar, a Fender bass) and headed for the shop. The shop in question was very small, maybe 10 feet wide by 20 to 25 feet deep. The owner, a Spanish émigré who had fled Generalissimo Franco's Spain, employed a part-time assistant. Arriving at the store, the owner recognized me and asked what I wanted. Pointing to the picture I said, "Can you make a guitar like this?" After studying the picture for a brief moment, he replied in a strong Spanish accent speaking a mixture of French and Spanish: "Creo que si, I have never made one like this before pero, no problema." He proceeded to spend a good half hour patiently explaining to me how he would have to reinforce the neck because of the tremendous pressure exercised by the strings but, he assured me he knew what to do. Next came the all important question: "How much?" "For you," he replied, "only 10,000 Francs." He never had any difficulty expressing himself in French when it came to numbers. Although that would essentially wipe out 80 to 85 percent of my life's savings, it was still a fraction of the cost of these fancy Fenders. I agreed.[2]

For the next few weeks, every time I walked by the small shop, I checked on his progress (no doubt to his annoyance). I watched him carve the body, then the neck. Next came the polish, the varnish, installing the electronics, and then one day, my red and black guitar was finally finished. But would it work? I bought the strings (which for a bass guitar are more like thick cables) and I began tightening them. The first problem I noticed was that regular tuning heads had been installed. The craftsman had told me that he could not find bigger tuning heads but he thought that would not be a problem. Well, he was almost right. I could certainly tune the guitar but my fingers felt the pain as I had to apply tremendous pressure on the tuning heads to turn those cables known as bass strings. However, that was the only inconvenience. The guitar worked perfectly and, in fact, it lasted some 30 years when, after much contemplation and some emotion, I finally got rid of it.

I have related the anecdote about the guitar because the interrelationship and exchange of information that occurred nearly 40 years ago when I placed an order for a customized guitar typifies the hundreds of millions of customer–supplier interfaces and transactions that occur every day throughout the world and have occurred ever since the first customer purchased from the first supplier. Vague customer requirements are routinely translated by craftsmen, machinists, engineers, industrialists, and businesspeople in general to produce products that somehow satisfy millions of customers. When I placed my order with the cabinet maker, I trusted his expertise. I did not interfere with his work by telling him how to set up his machine or how he should cut or polish the wood. I did not tell him

---

[1]Although the profession of wood craftsman is almost extinct, a few people in the United States and more overseas still preserve the craft.

[2]In those days, 1961, 10,000 francs was equivalent to about $US20.

where he should buy the wood nor what type of wood he should have used. I did not tell him how to make my guitar simply because I did not know and still do not know how to make a guitar, which is why I subcontracted for his expertise in the first place. Certainly, problems can occur, as was the case with the tuning heads of my guitar, but, in this instance, larger tuning heads were simply not available anywhere, and a substitute had to be found or the guitar could not be completed. Considering that the guitar was essentially a prototype, it is remarkable—and a credit to the experience, knowledge, and craftsmanship of the furniture maker—that the prototype was functional and durable.

To this day, thousands of suppliers throughout the world interact with their customers in the same fashion I did decades ago with the cabinetmaker. They receive an order for 10, 100, or a 1000 parts or assembly units. Based on customer requirements, which often can be as vague as the requirements for my bass guitar—for example, a sketch drawn on the back of a paper napkin—machine shops and workshops around the world can manufacture a prototype that may or may not be reviewed, inspected, and eventually approved by the customer. If not approved, adjustments are made until the customer is satisfied and orders the remaining 99 or 9999 parts or assembly units. It is true that this arrangement does not always guarantee reproducibility of the unit/product and or repeatability of the manufacturing process, but it is also true that in most cases the supplier/manufacturer/subcontractor does successfully produce identical parts or assembly units that are accepted by the customer.

Many companies and even whole industries, such as software developers, for example, are in the business of producing prototypes or constant variations of prototypes which are tested by customers—or a select group of customers—and updated with an unending flow of new and, supposedly, improved releases that are sure to eventually require the user to purchase more computer memory. Despite these obvious marketing strategies, the public does not seem to mind. On the contrary, many individuals cannot wait to purchase the latest, best upgrade, which will surely be obsolete within 12 months. Let us not forget the millions of products that are designed and manufactured daily with minimum or no input from customers/consumers. When we purchase a television, a radio, a refrigerator, a disk drive, a computer, or furniture, for example, the products we purchase were often conceived by engineers without our direct input (see Chapter 5 for an elaboration of this theme). This is not to say that engineers and management in general are totally oblivious to the needs and demands of customers/consumers (although, in some cases of poor design, one might think that is precisely the case). Certainly, in many cases, marketing surveys or field reports from field engineers are transmitted to design engineers who, after some initial resistance and denial, will eventually recognize that a design modification should be made. In the case of the 1997 Toyota, it is clear that engineers at Toyota listened to the so-called *voices* of the customers. These customer inputs resulted in many improvements that were introduced in the 1997 model.

Yet, there are major differences between the way Toyota, GM, Ford, and Chrysler interact as customers with their suppliers and the way I interacted with the cabinet maker. These customers do not hesitate to tell their suppliers or subcontractors how to run their business. They can control the production process of their suppliers not because it is better to do so (although they would probably argue that it is), but because they have the power to do so and lack the wisdom not to use it.

The Role of Power in Dictating Demands

Beginning in the 1950s and continuing to this day, political scientists and economists such as Gunnar Myrdal, Albert Hirschman, John Freidmann, Samir Amin, Gunder Frank, Johan Galtung, and many others described the economic relationship between the developed (center) and developing (periphery) world in the 1950s, 1960s, and 1970s as one characterized by unequal exchange. Unequal exchange occurs when developing countries trade low-priced raw material or agricultural goods for expensive finished goods (e.g., tractors and computers). This exchange leads to a trade deficit that is usually alleviated by bank loans, which are granted under harsh and often politically dangerous economic restrictions to the national economy (witness the activities of the International Monetary Fund in Korea, Indonesia, and Brazil in 1998 and 1999, for example).

This center-periphery economic model was eventually applied to the business world. Robert Averitt and John K. Galbraith, for example, described a dualism between dominant center firms (e.g., major corporations) and their subservient periphery firms (usually suppliers to these major corporations). In this dual economy, center or core firms use their market power to take advantage of firms on the periphery (suppliers). Averitt sees the business world not as a continuous series of small, medium-sized, large, and very large firms, but rather as a dual economic landscape dominated by a few large to very large powerful firms at the core or center of the economy and the rest consisting of smaller businesses.[3] Galbraith, who refined his theory during the course of writing several books, saw the American economy divided into two major sectors. In his *Economics and the Public Purpose*, Galbraith suggests that the two major sectors consist of a "planning sector" dominated by the largest 1000 firms and the "market system," which consists of the millions of smaller firms that are connected to the planning sector or otherwise supply the 1000 firms. Galbraith observed that firms belonging to the planning sector could impose, as a condition of conducting business, all sorts of conditions on smaller firms. They can impose prices, and at times they can demand costly quality requirements including certification (e.g., ISO 9000, ISO 14000, or QS 9000) requirements.[4]

Although a few authors have expressed mild skepticism at Averitt's dual economy (e.g., Hannan and Freeman; see below), others, such as Robert Frank, for example, share Averitt's observation:

---

[3]Robert T. Averitt, *The Dual Economy*, W. W. Norton & Co., New York, 1968.

[4]John K. Galbraith, *Economics and the Public Purpose*, Houghton Mifflin Company, New York, 1973. See particularly, pp. 42–44, 49–51, 201–202, and 252–260. Some of Galbraith's books include *The New Industrial State* (Houghton Mifflin Company, New York, 1967), *The Affluent Society* (Houghton Mifflin Company, New York, 1958), and *American Capitalism* (Houghton Mifflin Company, New York, 1952).

TABLE 1–1 Partial List of Mergers in the 4-Month Span from December 1997 to March 1998

Institutions	Value of Deal
Commercial Union and General Accident	$24.4 billion
Netherlands' ING bought Belgium's Banque Bruxelles Lambert	?
Sweden's Nordbanken merged with Merita of Finland (banks)	$10.6 billion
Morgan Stanley merged with Dean Witter; Bankers Trust bought Alex Brown; Travelers Group acquired Salomon Brothers; Merrill Lynch acquired Britain's Mercury Asset Management (financial institutions)	?
First American buys Deposit Guaranty (financial services)	$2.7 billion
Guinness merges with Grand Metropolitan (food and drink)	$39 billion
Ernst & Young merges with KPMG (accountancy firms and consulting)	?
Williams Company buys MAPCO (refining and pipelines)	$2.7 billion
Sonat (exploration) purchases Zilkha Energy	$1.04 billion
Allianz (German insurance) bids for AGF of France	$10.3 billion
Frito-Lay (Pepsi Cola) purchases Britain's United Biscuits	$440 million
Publicis (French advertising) bids for Chicago's True North	$700 million
Germany's Adidas (sportswear) buys Salomon (France)	$1.3 billion
Price Waterhouse and Coopers & Lybrand (consultancy) to merge	?
Britain's B.A.T. (tobacco and financial services) merges with Zurich Insurance	$37 billion
Anglo-Dutch Reed Elsevier merges with Dutch Wolters Kluwer	$20 billion
Lafarge (France cement maker) does hostile takeover of Britain Redland	$2.7 billion
Anglo-Dutch Unilever purchases Brazil's Kibon (ice cream)	$930 million
Chicago-based Ameritech purchases 42 percent of Tele Danmark	$3.2 billion
AT&T purchase of SBC Communications	$50 billion
Generale Des Eaux (water) buys 70 percent of Havas (publicity)	$6.5 billion
USA Waste Services purchases Waste Management	$13.0 billion
Alcoa acquires Alumax (aluminum)	$2.8 billion

Source: Reported by *The Economist* (Internet service); www.economist.com, 12/97–3/98.

The heightened competitive market which forces firms toward lean production may also be creating a whole new structure of free agency "winner-take-all" labor markets in which a handful of top performers walk away with the lion's share of total rewards.[5]

Indeed, the numerous billion-dollar mergers that have occurred since 1996 and continue to occur to this day throughout the world do not contradict Averitt's and Frank's observations (see Table 1–1). Large firms are getting larger and as they get larger they are not only able to further control their market but they are also increasingly able to place more demands on their suppliers (much as the military has been able to do during the past several decades).

[5]Robert Frank, "Talent and the Winner-Take-All Society," *The American Prospect*, Spring 1994.

As companies become larger, their economic dominance and power over so-called "peripheral" firms increase and so do their contractual demands or, rather, the contractual demands of their purchasing managers. As we shall see in a later chapter, the need to constantly serve their best and often most demanding customers can lead to unpleasant consequences for some firms. This unequal relationship is most clearly evident within the automotive sector, which, a few years ago, began to arbitrarily demand that all of its suppliers be certified to its prescriptive and demanding QS 9000 quality system requirements.

The Vendor–Vendee Relationship within the Automotive Sector: United States versus Japan

The business of achieving the right image of quality is very important for automotive assemblers. Within the automotive industry, the trend of hiring consultants to either pass an audit or achieve a high rating on crucial national surveys is certainly not new. For example, in the summer of 1994, Chrysler hired quality analysts to help its Neon division "score high on a crucial survey by marketing firm J. D. Power & Associates."[6] This tradition of hiring quality coaches to improve ratings is apparently also practiced by Ford and GM. Something seems very wrong with this scenario. Unfortunately for Chrysler, the hiring of a consultant to "pass" a survey and convince customers of the quality of its products was obviously not sufficient because by March 1995, the following headline was seen in newspapers: "Chrysler Halts Neon Production Again." The problem, a noise in the steering that does not affect the car's safety, was attributed to manufacturing and assembly operations.[7] Does the automotive industry truly believe that their quality management and tooling and equipment systems known as QS 9000 and TE 9000, respectively, will reduce the number of recalls? Moreover, can suppliers be held responsible for how cars are assembled? Finally, will QS 9000 and TE 9000, which are imposed on all suppliers to the automotive industry, help the big three become more efficient?[8]

James P. Womack and others explain that the evolution of the supplier–customer relationship in the United States and Japan has followed very different paths. The principal difference between the two systems is that, whereas the West still favors a dictatorial supplier–customer relationship, the Japanese, who improved the lean production system, emphasize a similar system euphemistically known as forced cooperation. Although the end result is intended to be the same, the nature and extent of control exercised by the customer over the supplier's means of production are not analogous and lead to some differences in the management of the customer–supplier relationship. According to Womack, Japanese car assemblers, which favor long-term relationships (which are usually

---

[6]See Micheline Maynard's "Quality Coaches," *USA Today*, August 8, 1994, p. 5B.

[7]Bill Vlasic in *USA Today*, March 16, 1995, p. 2B.

[8]The estimated number of cars built per worker in 1994 was 70.3 for Ford, 68.3 for Chrysler, and 57.1 for GM. Ironically, GM was the first to announce that all of its suppliers will be required to achieve QS 9000 registration. Could it be that GM officials truly believe that QS 9000 registration of their suppliers will help efficiency?

one sided), recognize that since the supplier is responsible for making the part, he should be allowed to, and is expected to, participate in the early stages of design. It is therefore the responsibility of the supplier to improve design and production techniques because the customer is willing to share the profits with him.[9] In the United States and Europe, where the adversarial/dictatorial relationship still dominates, these activities are unlikely to occur and are controlled by the customers' engineering department who in turn passes on requirements to the suppliers.

The means whereby Western and Japanese automotive customers evaluate their suppliers are very different. The Japanese constantly work with their suppliers to help them achieve a (supposedly) mutually agreed on price and delivery objectives. In the United States and Europe, manufacturers have instead developed, during the late 1980s and the 1990s, elaborate supplier surveys that, in essence, dictate what the supplier must do. The evaluation and rating of suppliers are certainly understandable when one recalls the long heritage of supplier evaluation and the rating system that originated with the U.S. military. Yet the use of such methods is certainly ironic because it presumes that the customer actually knows better than the supplier how to achieve the required quality for parts. But if this were true, why would the customer outsource the part in the first place?

Surely, automobile manufacturers such as Ford, Chrysler, and GM "transitioned" from manufacturers to assemblers during the 1950s because they realized that they could subcontract the parts, subcomponents, or subassemblies for *less* than it would cost them to manufacture. Of course, this did not guarantee that subcontractors could make parts with the required accuracy and/or reliability, and one wonders if the development and enforcement of a 14- or 20-page questionnaire (*à la* Ford Q-101) really guarantees quality.[10] Based on recent automotive recall statistics published by the National Highway Traffic Safety Administration, the answer would have to be "No."[11] Still, one cannot deny that the overall quality of American cars has improved during the late 1980s and

---

[9]The picture painted by Womack and others is a bit idealistic. For a very different analysis of the Japanese customer–supplier relationship, see Kuniyasu Sakai, "The Feudal World of Japanese Manufacturing," *Harvard Business Review*, November–December 1990, pp. 38–49. Sakai remarks that "...from the day a subcontractor accepts the first contract...it has given up its freedom." Sakai concludes by stating: "The way to create change is to get inside. Once inside, it is possible to offer the indentured servants an opportunity to escape, to work as they choose for whomever they choose ..." (pp. 40, 49). See also John E. Rehfeld, "What Working for a Japanese Company Taught Me," *Harvard Business Review*, November–December 1990, pp. 167–176. See also Tony Elger and Chris Smith (Eds.), *Global Japanization?*, Routledge, New York, 1994.

[10]I do not mean to suggest that other customers have not adopted the Japanese style of customer–supplier relationship. See, for example, Robert Ristelhueber, "Contract Manufacturing: Outsources Become Part of the Team," *Electronic Business Buyer*, August 1994, pp. 48–56, where the author covers many of the issues of so-called lean manufacturers as explained in Womack's book.

[11]Recent statistics show that the total number of vehicle recalls has increased steadily from approximately 6 million vehicles in 1990 to 11.1 million vehicles in 1993 (statistics reproduced in James R. Healey, "Quality Woes Are Spreading Like a Rash," *USA Today*, March 18, 1994, p. B1). Original source: Auto Service Monitor, National Highway Traffic Safety Administration. See also Jayne O'Donnell, "Seat Belt Study Finds Hazard," *USA Today*, March 18, 1994, p. B1.

1990s. And yet the Big Three, Ford, GM, and Chrysler, agreed in mid-1994 to continue the tradition by developing and imposing on their suppliers yet another system known as QS 9000, which appends the requirements of the automotive industry to those of the internationally recognized ISO 9001 quality assurance model! (See Chapter 12 for a description of the ISO 9000 series of standards.)

Naturally, third-party registrars were quite excited at the prospect of having to register the thousands of suppliers to the automotive industry. Within the automotive industry, the business of regulating suppliers is not likely to end and will generate more revenues for the "experts" that are lining up to conduct the lucrative business of official interpreters of quality as ordained by QS 9000.[12] Naturally, to correctly interpret QS 9000, auditors have to be formally trained to become approved or certified QS 9000 auditors. Potential auditors can achieve that requirement by completing yet another course, especially designed for the "rigorous" needs of the automotive industry. The price: a mere $500 (1997) per course.

Dual Economy in the World of International Standards

The dualistic model is also present in the world of international standards, particularly the ISO 9000 series of quality systems. The various committees that produce these standards tend to be dominated by individuals who generally can only think in terms of a big company mind-set. As the standards become longer, more prescriptive, and more demanding, it is evident that only large companies, equipped with a bureaucracy and resources, will be able to satisfy the unending and increasingly prescriptive demands of international quality and environmental standards such as the ISO 9000 and ISO 14000 series. Similar comments could be made about the automotive's QS 9000 standard. As Michael T. Hannan and John Freeman have noted:

> *All adaptation theories agree that the largest, oldest, and most powerful organizations have superior capacities for adapting to environmental circumstances. Size and power enable organizations to create specialized units to deal with emerging environmental problems. More important, these characteristics convey a capacity to intercede in the environment and to forestall or direct change.*[13]

The economic relationship between certain powerful customers and their suppliers can be viewed as a center-periphery type relationship. Powerful organiza-

---

[12]This assumes there are 3000 to 6000 suppliers to the automotive industry and that an average audit will cost a conservative $15,000. The potential revenue for registrars is anywhere between $45 to $90 million just for audits. However, because most registrars like to add preassessments, an additional $9 to $22.5 million can be achieved. Because the certificate is only good for 3 years, total potential revenues could reach $55 to $112 million dollars—to be renewed every 3 years! This does not include consulting costs and implementation costs.

[13]Michael T. Hannan and John Freeman, *Organizational Ecology*, Harvard University Press, Cambridge, MA, 1989, p. 12.

tions such as the big three (QS 9000), Boeing (D1 9000), aerospace manufacturers (AS 9000), GE and Motorola with their six-sigma program, the German automotive industry with its VDA 6.1, and many other large corporations as well as U.S. and foreign government agencies routinely dictate to their suppliers how they must conduct their business. This they achieve by requiring suppliers to implement a prescriptive quality management system, which interferes with the suppliers' operations. Other examples of intrusive/prescriptive regulations would include the FDA's Medical Device Regulations (21 CFR Part 820, April 1998) and countless other national and international (ISO) standards.

This philosophy of control guided by the "more is better" mentality has even influenced the ISO 9001-2000 standard. These intrusive models for quality assurance can only be imposed because certain "customers," or their self-appointed representatives, have the power to dictate specific terms and conditions to less powerful businesses. This ability to impose requirements irrespective of cost, as was practiced by the military in the 1950s and even earlier (see Part III), has now been adopted and modified by some ISO 9000 registrars, the very organizations that have emerged as the self-appointed guardians and official interpreters of the ISO 9000 series of standards. When combined with bureaucratic narrow-mindedness and meticulousness, power can sometimes produce severe and costly discomfort.

CHAPTER 2

On Registrars and Bureaucratic Power

A bureaucratic organization is an organization which cannot take corrective action, based on its errors.

—*Michel Crozier*[1]

Paul Tabori, author of an unusual book about stupidity, comments: "There is a Turkish proverb that says: 'If Allah gives you authority, He will give you the brains to go with it.' Like many proverbs, this one is both dangerous and false. As far as bureaucracy is concerned, the acquisition of authority more often than not leads to a loss of brains, to an atrophying of the mind, to a chronic state of stupidity."[2] Most people who have had to deal with bureaucracies would probably agree with Paul Tabori. And yet the truth is that bureaucrats are not necessarily stupid or any more stupid than any other administrator; however, they must act within the constraints of a complex system of rules and regulations that often forces them to act in a silly, if not occasionally stupid, way.

In his book *Bureaucracy: What Government Agencies Do and Why They Do It*, James Q. Wilson observes that the difference between private and public bureaucracies has to do with constraints. Public organizations must operate within political constraints imposed by state or local officials. Moreover, as public demand increases, public organizations cannot arbitrarily raise their revenues to better serve "customers." Instead, political appointees who reside in the nation's capital arbitrarily control and adjust the organizations' budgets as they see fit. Public organizations must also ensure that they represent everyone fairly and, as Wilson explains:

---

[1]Michel Crozier, *Le Phenomene Bureaucratique*, Editions du Seuil, Paris, 1963, p. 229. "Une organisation bureaucratique serait une organisation qui n'arrive pas à se corriger en fonction de ses erreurs."

[2]Paul Tabori, *The Natural History of Stupidity*, Barnes and Noble Books, New York, 1993, p. 100.

To a much greater extent than is true of private bureaucracies, government agencies (1) cannot lawfully retain and devote to the private benefit of their members the earnings of the organization, (2) cannot allocate the factors of production in accordance with the preferences of the organization's administrators, and (3) must serve goals not of the organization's own choosing. . . . As a result, government management tends to be driven by the constraints on the organization, not the tasks of the organization."[3]

Thus, whereas business focuses on the bottom line, government management focuses on the top line (that is, constraints.)

Constraints and Absurdity

Within the United States and no doubt most other nations, examples of dubious decisions caused by constraints abound. In an increasing number of countries, the subject of environmental protection and the associated plethora of environmental regulations or laws continues to gain momentum. In the United States where environmental laws have been in effect for almost 30 years, examples of absurd governmental decisions regarding the environment abound.[4] One particular case originated in San Bernardino County in Southern California, and is centered around the controversial Federal Endangered Species Act. This act is designed to protect thousands of supposedly endangered species. The act is controversial on at least two counts: First, not everyone agrees as to what is or is not an endangered species and, second, once a real estate developer (or any citizen) wants to develop a region or property where an endangered species allegedly lives, environmental bureaucracy takes over. Because the developer or property owner must now work with government biologists to ensure that the habitat is not substantially disturbed so as to have a negative impact on the viability of the (endangered) species, development costs may double or triple and in some cases, development may be permanently halted.

In San Bernardino County, California, one of the many endangered species protected by the Endangered Species Act is the Delhi Sands *fly*. When developers wanted to build a hospital, local environmentalist groups fiercely objected because of the potential threat to *the fly*. County officials sided with the developers and refused to enforce the Endangered Species Act. The case was scheduled to go to the U.S. Supreme Court for a ruling but, unfortunately for county officials and the developers, the Supreme Court refused (perhaps wisely) to take the case. County officials had no alternative but to enforce the Endangered Species Act. The hospital lost and the fly won, which means that if a hospital is to be built, land developers must work with government biologists to ensure that the fly's environment is not disturbed. This is a classic case of one bureaucracy

---

[3]James Q. Wilson, *Bureaucracy: What Government Agencies Do and Why They Do It*, Basic Books, New York, 1989, p. 115.

[4]To protect an endangered fly, the Department of Fish and Wildlife once suggested that speed on a freeway (Interstate 10) between two cities be reduced to 20 miles per hour!

with its own set of constraints (San Bernardino County) facing a bigger bureaucracy with a larger set of constraints (the Environmental Protection Agency [EPA] whose decisions impact the whole country). In this instance, the larger set of constraints won the day which means in today's rationality that a fly is more important than a hospital.[5]

The second example relating to the limitations of procedures is equally absurd. A teacher who suffers from a rare skin disorder that prevents him from having his fingerprints taken was refused a teaching job because California State laws (the ultimate procedure) require that all teachers provide fingerprints so any criminal records can be investigated. After several months of indecision on the part of state bureaucrats, the teacher decided to find another job.[6]

This inability on the part of rigid bureaucracies to address unique requests/needs had been observed decades ago by the sociologist Robert Merton. Confirming Merton's observation and anticipating the works of sociologist Alvin Tofler and quality guru Tom Peters by several decades, the French sociologist Michel Crozier observed that no organization can survive unless it becomes flexible and capable of adapting itself. To achieve those objectives of flexibility and adaptation, Crozier suggests that organizations must trust the initiative and ability to invent found in certain individuals or groups.[7] But this is precisely what public organizations are not very good at doing.

Types of Organizations

Not all organizations are created equal nor do they operate within the same set of constraints. Wilson proposes four types of organizations, as listed in Table 2–1. Wilson's differentiation is valuable because it recognizes that no organization or government institution is created equal. (A point that has been ignored by the ISO 9000 committees who seem to believe that all organizations have equal resources.) Note: The ISO 9000 series of standards is discussed in Chapter 12.

Of the four types of organizations described by Wilson, the most interesting are the so-called "procedural and craft organizations" because of their direct connection with the ISO 9000 phenomenon. As Wilson explains, in procedural organizations management becomes means oriented. *How* the operators go about

---

[5]"Supreme Court Swats Fly's Foes," *The Press-Enterprise*, June 23, 1998, pp. 1, 6. California, which unfortunately for developers is well known for its plethora of environmental laws and a strong pro-environment lobby, is currently experiencing several legal battles regarding the protection of so-called endangered species. Other species which have recently frustrated developers include a rodent known as the kangaroo rat (or k-rat) and the Quino butterfly. The Quino butterfly spends most of its time below ground and only flies for 4 to 8 weeks before it dies. It feeds on two native grassland plants. Any landowner who has on his or her property these two native plants cannot develop the property until proper environmental impact surveys have been conducted. These surveys cost at least $3000 plus additional inspection cost of up to $1000. "Butterfly Panics Area Land Sellers," *The Press-Enterprise*, March 18, 1998, p. B1, B5, see also "County Builds Verbal Traps for Its K-Rats," *The Press-Enterprise*, February 25, 1998, p. B6a.

[6]"Teacher Drops Job Bid over Fingerprint Rule," *The Press-Enterprise*, June 13, 1998.

[7]Crozier, *Le Phenomene Bureaucratique*, p. 228.

TABLE 2–1 Four Types of Organizations

Type of Organization	Definition and Examples
Production organizations	Organization where both outputs or work and outcomes are observable (and thus measurable). *Examples:* Internal Revenue System, U.S. Postal Service, Social Security Administration
Procedural organizations	Organizations where managers can observe what their subordinates are doing but not the outcome that results from those efforts. *Examples:* Occupational Safety and Health Administration, the armed forces
Craft organizations	Organizations that consists of operators whose activities are hard to observe but whose outcomes are relatively easy to evaluate. *Examples:* Army Corps of Engineers, U.S. Forest Service
Coping organizations	Organizations where neither the output nor the outcome of their operators can be observed. *Examples:* schools and universities

Source: Adapted from James Q. Wilson, *Bureaucracy: What Government Agencies Do and Why They Do It,* Basic Books, New York, pp. 159–171.

their jobs is more important than whether doing those jobs produces the desired outcomes.[8] For many companies that have implemented one of the ISO 9000 standards, this procedural phenomenon is probably the fundamental flaw of the ISO 9000 registration process. In many cases the emphasis on following procedures can be directly attributable to third-party auditors and the organization they represent, namely, ISO 9000 registrars (see also Chapters 10, 12, 14, and 15).

Registrars as a Mixture of Craft and Procedural Bureaucracies

Many registrars (i.e., third-party organizations that grant ISO 9000 certification) are the epitome of a blend between procedural and craft organizations. The procedural aspect of registrars is easy to understand. After all, prior to issuing an ISO 9000 certificate, registrars must ensure that an organization has implemented a variety of procedures as called for by the international standards. However, because the international standards do not specify how many procedures must be written or how long and detailed they should be, many registrars have developed their own set of interpretive guidelines to supposedly assist their auditors in determining whether or not an organization has adequately addressed the intent of the standards. In many cases, these guidelines have evolved into a set of standard operating procedures (in this case standard operating auditing procedures) known as checklists which are nothing more than a rephrasing of the standard with some additional requirements.

---

[8]Ibid., p. 164.

The craft aspect of registrars is also easy to comprehend when one understands the certification process. The ISO 9000 certification process consists of having one or more auditors visit a company for 2 or more days to assess whether or not the organization is compliant with the set of requirements listed in the standard. Once the audit is completed, the auditor(s) must complete a report that varies in length depending on the registrar (but can be up to 20 or more pages). The report is submitted to a group of experts for review. Oddly enough, it is these experts—*who have not audited the facility and who often may not have even seen any documents*—who have the authority to determine whether or not an organization is worthy of certification. Such operational behaviors are typical of craft organizations. In this case, the activity (auditing) is not even witnessed by the group of experts. (This function is delegated to the auditor who is responsible only for the audit process but not the final decision!) All the experts can rely on to reach a decision is the auditor's outcome, namely, the report submitted by the auditor—which usually includes a recommendation for or against certification.

Because the group of experts who has the authority to grant certification is not present during the audit one is left with an unusual, if not absurd, scenario. To control the audit process, registrars use a checklist and a set of specific forms to control as much as possible what their auditors must do during an audit. Rather than trust the auditor(s), the group of experts have, on occasion, devised a complicated set of rules to monitor the auditor (who is an expert in his or her own right).

Having audited for a European registrar whose managers love to develop detailed and cumbersome procedures, I can attest that this unusual set of circumstances has often led to situations where one expert who operates from a European capital known by his compatriots as the "all knowing center of the universe," refuses to grant certification because of some peculiar interpretation of the standard! And yet, European registrars certainly do not have a monopoly on whimsical decisions (similar stories could be told of other European or American registrars); arbitrary interpretations rendered by so-called experts have been and will continue to be the Achilles' heel of any certification program for many decades.

Virtual ISO Certification: Guaranteed, Cheap, and Easy

The business of ISO 9000 registration has now become so competitive that some registrars pride themselves on guaranteeing fast and easy implementation. In some cases one only needs to purchase a CD or a software package, change a few names, ensure that there is enough supporting documentation to match what is promised in the software package, go through one or two rounds of internal audits, and—presto!—certification follows within a very few short months. For around $1300 a day, some organizations will provide clients with a $250- to $300-a-day consultant who will assist a client with a rapid (90-day) ISO implementation. The "guaranteed" implementation is then validated a few short weeks later by the registration side of the same organization. Some audits only last 1 day, just long enough to walk or, rather, run through a plant and issue the proper certificate.

I would expect that within the next couple of years ISO 9000 certification will be available via the Internet. All one would have to do is log in to a web site and select among the many options available the appropriate hyperlink text. Once connected to a secure site, one would have to fill out a couple of electronic forms, including a credit card number. To legitimize the process, an Internet videoconference would be scheduled. During the (video) session, the auditor, operating from the comfort of her home would obtain from a company representative a password and "audit" the facility by simply accessing various files found on the company's intranet network or perhaps even directly from a web site. If the auditor wishes to interview an employee she could do so using a video-conference system. Satisfied with the audit, the auditor would then complete her report on line and send it to the appropriate committee for review (this step is actually redundant but still required) and the registrar's "central office" would, after credit card approval has been verified, issue (via e-mail of course) an ISO 9000 or ISO 14000 or whatever else certificate. Welcome to the world of virtual auditing.

Part II

The Limits of Quality: Essays on a Separate Reality

The fact of the matter is that the "real world" is to a large extent unconsciously built upon the language habits of the group. No two languages are ever sufficiently similar to be considered as representing the same reality. The worlds in which different societies live are distinct worlds, not merely the same world with different labels attached. . . . We see and hear and otherwise experience very largely as we do because the language habits of our community predispose certain choices of interpretation.

—Edward Sapir (1929)[1]

[1]Edward Sapir, "The Status of Linguistics as a Science," in D. G. Mandelbaum (Ed.), *Culture, Language and Personality*, University of California Press, Berkeley, 1958, p. 69.

CHAPTER 3

Thoughts on the Relativity of Quality

He does not grasp that I do not care too much about quality. Look, I am used to things falling apart, I explained to him, I don't feel guilty when I buy them, because I am not throwing money away by paying too much.

—Slavenka Drakulic[1]

In the summer of 1996 a friend of mine whom I knew from my days at Boeing in Seattle, Washington, decided to quit Boeing. The fact that anyone should resign from a job is not particularly unusual; however, what was rather unusual about my friend's decision is that at age 38, Mike decided that it was time for him to travel around the world. Having made his decision, he packed up some of his belongings in a bag and, accompanied by his girlfriend, took off from Seattle headed for somewhere in Asia. As of late 1998 they were still traveling and had some vague expectation of being back in the states perhaps later that year!

Every 4 to 5 months I would receive a letter or e-mail from Mike: Australia, Indonesia, Cook Islands, New Zealand, Borneo, Thailand, Malaysia, Vietnam, Nepal, India, Afghanistan, Zimbabwe, Dar es-Salem, and so on. I always enjoyed reading about their adventures and occasional frightening mishaps. Because Mike used to be a quality engineer at Boeing, he sometimes described in his letters his impressions of what is perceived as quality or quality of service in these remote parts of the world. His many stories could easily comprise a chapter and would horrify anyone. In one of his letters where he described water quality problems in India, Mike wrote: "We have just finished reading an article in *India Today* about all the bottled water manufacturers here in India that produce the only water really 'thought' safe to drink here. The tests show that *none* of it is within safe acceptable BIS (Bureau of Indian Standards) or PFA Act (Prevention of Food Adulteration) levels! The editor stated that 'quality is usually by accident, not design.'"[2] What I find interesting about Mike's comments is not only that the

---

[1]Slavenka Drakulic, *Café Europa: Life After Communism*, Penguin Books Ltd., London, 1999, p. 73.
[2]Personal correspondence.

water is not within acceptable limits but that apparently even though legislation exists it carries little if any weight because the BIS does not seem to have adequate resources to do anything about it.[3] Yet, before pointing a critical finger at what seems to be a problem so commonly found in "developing" countries, one should realize that such problems are not limited to the developing world. In a September 4, 1998, article titled "State to Monitor Vending Machine Water," one learns that in Los Angeles County, 93 percent of all vending machines "had bacteria levels up to 163 times as high as in domestic tap water . . . 38 percent of the machines checked were failing to remove common organic compounds—including trihalomethanes, a chlorine by-product that in high levels has been linked to increased incidents of miscarriage and 62 percent (of machines that were supposed to dispense purified water), contained dissolved solids—minerals—exceeding the state limit."[4]

When I once wrote Mike that someone had contacted me to conduct some quality training in Nepal related to the ISO 9000 series of standards, Mike wrote back: "I can't even imagine any entity in Nepal being ready to encompass strict quality standards, let alone ISO 9000! It will most likely be a very difficult assignment, Jim."

His letter, which was mailed from India, was dated December 27, 1997, and arrived in my mailbox on July 2, 1998! But who is to say that a 6-month delivery time is unacceptable? Patience and concepts of time are probably very different in India and it may well be (although I doubt it), that 6 months' delivery for a letter is perhaps considered within acceptable norms in India. As every multinational who has invested abroad in search of ever cheaper labor continues to discover, when the time comes to deal with local suppliers, concepts of quality and on-time delivery are indeed relative.

The Zen of Quality

At least two major connotations are associated with the word "quality." The first, generally understood by quality professionals, refers to a measurable quality that is characterized by repeatability of processes and/or uniformity of well-defined measurable characteristics of a product or service. In terms of statistical measures, this "quality" is often assessed in terms of minimum variability.

The other aspect to quality, rarely if ever mentioned by quality professionals, is much more difficult if not impossible to measure because it is often associated with craft or craftsmanship as described in the opening pages of this book. And yet, although it cannot be quantified, most people would recognize this metaphysical "quality" when they see, feel, or taste it. Robert Pirsig, in his popular

---

[3]A similar problem exists with the enforcement of environmental laws in developing countries. Many developing countries have written or borrowed countless laws; however, they do not have the resources or the cultural mind-set to enforce them. See James Lamprecht, *ISO 14000: Issues and Implementation Guidelines for Responsible Environmental Management*, AMACOM, New York, 1997, pp. 34–36.

[4]"State to Monitor Vending Machine Water," *The Press-Enterprise*, September, 4, 1998, p. A3.

Zen and the Art of Motorcycle Maintenance, spends the greatest part of his book explaining what this indefinable craftlike quality is. How Pirsig came to write about the metaphysics of quality (moq) is interesting. Pirsig used to teach English composition at a university in Montana in 1959. One day, his secretary Sarah casually commented to Pirsig: "I hope you are teaching Quality to your students."[5] His immediate reaction was to confess that he did not know what quality was: "What the hell is Quality? What *is* it?"[6] That simple question was to lead Pirsig on a long and at times, tortuous intellectual search. It is important to understand that Pirsig is not talking about measurable quality. As Pirsig explains using his alter ego Phadreus:

The Quality that he (Phadreus) and the students had been seeing in the class-room was completely different from the qualities of color or heat or hardness observed in the laboratory. Those physical properties were all measurable with instruments. His Quality—"excellence," "worth," "goodness"—was not a physical property and was not measurable. . . . To arrive at this Quality requires a somewhat different procedure from the Step 1, Step 2, Step 3 instructions that accompany a dualistic technology.[7]

Yet, Pirsig's investigation into the realm of this moq is well worth exploring because it reveals some interesting observations of value to anyone involved with the search for quality.

Quality, Pirsig proposes, is a characteristic of thought that cannot be defined. "But even though Quality cannot be defined, *you know what Quality is!*" Pirsig recounts a story when he was teaching. He would ask his students to read and rank several essays in terms of quality. To his surprise the students' rankings—with only one or two exceptions—would always agree with his own evaluation of what were good or bad essays. Although everyone would agree as to what were good essays, no one could define the characteristics of a good essay, let alone offer a definition of what constituted a "quality essay."

Pirsig also recognizes that quality can be relative (see below). "People," Pirsig observes, "differ about Quality, not because Quality is different, but because people are different in terms of experience."[8] This rather obvious observation seems to have been ignored by some quality professionals who persist in wanting to standardize quality by standardizing and exporting quality management systems, management styles, or other practices. In essence, their effort is to ultimately standardize manufacturing practices and eventually the international workforce. If carried to extreme, this standardization process could, in the long run, lead to Aldous Huxley's brave new world, a place of "Standard men and women; in uniform batches. The whole of a small factory staffed with the products of a single bokanovskified egg."[9]

---

[5]Robert Pirsig, *The Zen of Motorcycle Maintenance*, William Morrow, New York, 1999, p. 180.

[6]Ibid., pp. 183–184.

[7]Ibid., pp. 231, 292.

[8]Ibid., p. 250.

[9]Aldous Huxley, *Brave New World*, Harper Perennial Classics Editions, Harper Collins, New York, 1998, p. 7.

Is There a Universal Approach to Management?

I would suspect that most, if not all, supporters of standardization (i.e., the popular ISO 9000, the international standard for quality management systems) would answer the above question in the affirmative. To them, a raison d'être of ISO 9000 certification is to introduce a universal principle of management. Proponents of an ISO 9000 "universalism" would argue that in a perfect world, universal quality would at long last be achieved when all companies throughout the world are ISO 9001-2000 certified. But is that really desirable or even possible? Can an international standard be uniformly adopted across culture without experiencing some adaptation?

Some years ago, the Dutch sociologist Geert Hofstede undertook an extensive study that established quantitatively that there were, not surprisingly, differences in work habits across cultures.[10] Expanding on Hofstede's work, another Dutch researcher by the name of Fons Trompenaars conducted, over a period of several years, an extensive quantitative analysis with the aim of better understanding, describing, and classifying national and corporate cultures. Among the many valuable conclusions reported by Trompenaars and Charles Hampden-Turner, I comment on only one aspect of the authors' research that is of significance to the worldwide ISO 9000 certification movement.

Trompenaars and Hampden-Turner recognize four broad types of corporate cultures[11]:

1. The family
2. The Eiffel tower
3. The guided missile
4. The incubator.

One characteristic of the family corporate culture is that it tends to be characterized by personal and hierarchical relationships, which usually lead to a power-oriented corporate culture. *Who* is doing something is more important than *what* is being done. Training is not used to challenge authority but rather to perpetuate it. Within the family corporate model, efficiency (doing things right) is not as important as effectiveness (doing the right thing).

Example of countries where a family corporate culture is likely to be found include Turkey, Venezuela, India, Malaysia, Mexico, and, to a lesser extent, France, Spain, Argentina, and Greece. (The authors provide a more complete country list across several cultural dimensions.) Companies belonging to the family corporate model usually experience difficulties implementing the ISO 9000 model.

---

[10]Geert Hofstede, *Culture's Consequences: International Differences in Work-Related Values*, Sage Publications, Beverly Hills, CA, 1980. See also G. Hofstede, B. Neuijen, D. Ohayv, and G. Sanders, "Measuring Organizational Cultures: A Qualitative/Quantitative Study Across Twenty Cases," *Administrative Sciences Quarterly*, 35, 1990, pp. 286–316.

[11]Fons Tromprenaars and Charles Hampden-Turner, *Riding the Waves of Culture: Understanding Cultural Diversity in Global Business*, 2nd ed., McGraw-Hill, New York, 1998, pp. 162–185.

The Eiffel tower corporate culture is very different from the family corporate environment. Within the Eiffel tower corporate structure, relationships are specific and status is ascribed. An ascribed status is attributed to you by birth, kinship, gender, or age, but also by your connections and your educational record. In France, Japan, or Mexico, for example, the name of the university from which one graduates is very important. Status based on achievement, in which one is more likely to be judged on recent accomplishments or a well-established performance record, is a quality more likely to be found in the United States. In a society that favors achievement, *what* you studied is more important than *where* you studied (ascriptive). Countries where stereotypic Eiffel tower corporate cultures are likely to be found include Germany and Austria.

"Change in the Eiffel Tower is effected through changing rules. . . . Manuals must be rewritten, procedures changed, job descriptions altered, promotions reconsidered, qualifications reassessed."[12] As Trompenaars and Hampden-Turner observe, the Eiffel tower does not adapt well to turbulent environments, nor does it relate well with family corporate culture. These observations would indicate that an ISO 9000 management system is better suited for an Eiffel tower corporate culture.

The guided missile corporate culture characterized by U.S. and Dutch corporations, for example, resembles the Eiffel tower in that it is impersonal and task oriented. The culture is neutral. Learning includes "getting on" with people, being practical rather than theoretical, and a problem-centered rather than a discipline-centered approach. Loyalties to professions and projects are greater than loyalties to the company. As can be expected, companies favoring this corporate culture are not as comfortable with the structure of the ISO 9000 system as companies favoring the Eiffel tower model.

Finally, the incubator culture, usually found in high-tech companies (Microsoft, Silicon Valley companies, etc.), is characterized by self-expression and self-fulfillment. These are creative organizations. There is intense emotional commitment but not necessarily toward other workers; instead it is directed at the work being undertaken. The incubator culture has been compared to a jazz band where improvisation is the essence. Because the customer often has not defined a target nor does he have any idea as to what the technical specifications need to be, problems can be easily and constantly redefined. Leadership is achieved, not ascribed. Incubator cultures are also generally not receptive to the rigor and structure imposed by the ISO 9000 system.

Given the broad range of corporate cultures that has been identified by Trompenaars and Hampden-Turner, can one really expect the monolithic structure of the ISO 9000 management system to be applied with equal success throughout the world? Based on this author's experience with ISO 9000 implementation during the 1990s, the ISO 9000 model for quality management—which is influenced by the Eiffel tower corporate mentality—is generally not well received or comprehended within a family or incubator corporate culture.

---

[12]Ibid., p. 174.

Are There Universal Principles of Quality?

There *seems* to be a fundamental or universal level of appreciation or acceptance for certain things. In the case of many movies produced in Hollywood, it would appear that movie producers have not only mastered the art of computerized special effects but also know how to appeal to the lowest common denominator, perhaps the source of universal taste. Naturally, one has to exercise caution when writing about "fundamental" or "universal" tastes because one cannot ignore the important role played by international marketing and advertising agencies as they relentlessly try to influence, standardize, or otherwise manipulate global wants and desires. At a minimum, one can suggest that although national differences in skepticism vis-à-vis advertising campaigns must exist, most people throughout the world seem to respond or at least be influenced or manipulated by international marketing agencies in a similar fashion (hence the popularity of certain international superstars who have made good use of the powerful combination of international marketing and multinationals such as SONY, for example).

People across the world (at least the Western world) seem to be able to appreciate or be convinced to appreciate similar musical and motion picture values and, to some extent, (fast) foods. In the case of rock music, the appeal definitely goes beyond the Western world and reaches, with various success, the far corners of Africa and Asia. When American movies are shown in Europe (and other parts of the world), an American way of life, and hence part of its culture (or rather, Hollywood's version of it), is exported and presented to millions of viewers. When the process is repeated every week, for decades, a process of acculturation may take place. In some countries the potential for acculturation is considered a threat. Faced with this audiovisual cultural invasion, French cultural ministers have, over the years, often complained about this American form of cultural colonialism and have tried in vain, to impose quotas on American movies.[13]

If the existence of some form of universal appreciation can generally be accepted at least for some products, can we also assume that universal values of quality exist such as quality of service, quality of products, quality of management? Moreover, when one tries to standardize quality (management) principles or instill a culture of quality developed in another country or continent, can we then speak of a process of acculturation with respect to quality principles? Can we ignore or dismiss the important role of the receiving culture, which may modify or otherwise influence these "imported" quality values?

For example, when multinational corporations invest overseas, can they willingly or inadvertently introduce (or transfer) their (foreign) work culture-ethic and culture of quality? Certainly, there is no doubt that principles of quality control have been successfully "transplanted" overseas. Having visited numerous assembly plants operated by multinationals in Europe, Central America, and Latin America, I can attest that a control chart prepared in Spain or Brazil looks the same as in the United States. Operators, irrespective of their nationality, ask similar questions and make similar errors or recommendations when it comes to

---

[13]See Richard Pells, *Not Like Us*, Basic Books, New York, 1997, pp. 214–220.

collecting or interpreting data. Even though quality managers in Europe, Brazil, or Mexico work in management–labor environments governed by different national laws, they are faced with similar quality-related problems, share similar concerns, and often suggest similar (but not identical) solutions.

To what can one attribute these similarities? Are there universal principles of quality that are so obvious as to be unconditionally accepted by all managers throughout the world? Or are other forces at play? I would propose that although people around the world may ascribe to certain universal values of quality of product or quality of service, part of the homogenization of these ideas and philosophies about quality, at least within the managerial world of enterprises, is in part due to a pattern of cultural dominance. I am specifically referring to the translation of quality concepts and principles from English to other languages.

Can Quality Be Translated?

When was the last time you read a book about quality or quality-related issues such as management that was an English translation of a French, Spanish, Portuguese, or German work? With the exception of a handful of books originally published in Japanese, I know of no such books that have ever been translated because the translation of quality or managerial concepts, theories, or principles is overwhelmingly unidirectional: from English to other languages (but not the other way around). Similarly, one should note that the origin of the ISO 9000 series of quality assurance standards came from the English-speaking world and not from the Spanish-, Arabic-, or Chinese-speaking world.

Whenever I travel to the francophone province of Quebec, I always try to visit bookstores. Two things have always surprised me about bookstores, no matter how large, in Quebec (and also France, Spain, and Italy, for example). The first, is that it is always difficult to find the "management section," where most books on quality are usually found. One always finds the section for essays, novels, political or social commentaries, or philosophy but not management. To find books dealing with management or quality, one often has to go to specialized bookstores, but even then, the section is often small.

This unidirectional flow of information as to what general principles of quality should be often results in ridiculous scenarios. In Mexico, for example, Mexican consultants are often unfairly perceived by Mexican companies as being less competent than foreign consultants. In some cases, Mexican firms will not hesitate to pay as much as 5 to 10 times the daily rate of Mexican consultants just to hire foreign expertise. And yet my association with a few Mexican consultants has convinced me that there are in Mexico many excellent consultants who certainly are as good as any foreign consultants. This unjustified prejudice against their own kind and reverence for anything foreign is unfortunate because it invites Mexicans to unconditionally accept criticism of what they do best. If there is one thing Mexicans are known for, it is their friendliness and hospitality toward foreigners. When it comes to quality of service as defined by the criteria of friendliness and even efficiency, I have had nothing but pleasant experiences in Mexico. I would even suggest that American hotels should benchmark some Mexican hotels (and not vice versa) to learn how to better serve their customers. And yet, some American consultants would not hesitate to suggest that when it comes to customer service, Mexicans better learn from Americans. What is stranger still is that Mexicans generally accept such criticism.[14]

The second surprising aspect of bookstores in Quebec, Europe (excluding the United Kingdom), or Latin America (with the exception of Brazil and possibly Buenos Aires, Argentina) is that when you find books on management or quality they are invariably translations of books written by American authors. Books about empowerment, reengineering, ISO 9000, ISO 14000, leadership, activity-based accounting, statistical techniques, paradigm shifts, and so on have all been translated from English. All of the latest quality and/or managerial fads are usually available within a few months of their original release in the United States (at least in Brazil). Of course, the ratio of American translations to native authors will vary from country to country, but, still, most management books or books covering a broad range of topics related to quality are translations. The flow of ideas, concepts, and theories about what quality should be (at least for the next 12 to 18 months) is coming from the United States. No wonder Spanish, French, Brazilian, and other managers sound like they have read the latest book by management guru Tom Peters; they have and because many foreign managers can read English, they probably have read Peters and others like him in the original English version.

Once quality principles have been translated into several languages, can one ignore the culture within which these translated quality principles are supposed to operate? It is one thing to have a conversation with a quality professional from France, Brazil, or Spain, but it is a different thing entirely to see how the boss–employee relationships operate within the context of a particular industrial culture which itself is a subset of the national culture.

When my wife and I had the good fortune to live in Paris in the early 1990s, we occasionally noticed that, in our no doubt prejudiced opinion, shoppers did not seem to experience the best customer service or customer–employee relations in the world. On more than one occasion I recall seeing unoccupied salespersons stand next to a ringing phone never attempting to answer the phone. Parisian salespersons *appeared* to be distant, occasionally gloomy, or even unfriendly. Why is that so? Social conditions, French labor laws, and labor relations in France in the early 1990s (I no longer know what the conditions are today

---

[14]See, for example, Claire Poole's "Service Doctor," *Mexico Business*, January–February 1997, p. 3133. For an excellent description of why Mexicans may still be dependent on foreigners, see Mauro Rodriguez Estrada and Patricia Ramirez Buenda, *Psicología de Mexico en el Trabajo*, McGraw-Hill, Mexico, 1996. What is odd about Rodriguez and Ramirez's book is that in their attempt to inform their readers (Mexicans) of their psychological flaws and low self-esteem vis-à-vis foreigners, the authors offer a particularly harsh criticism of Mexicans in general. Other classic works would include Wallace Thompson's *The Mexican Mind: A Study of National Psychology*, Little, Brown and Company, Boston, 1922; Samuel Ramos' *Profile of Man and Culture in Mexico*, University of Texas Press, Austin, 1962 (first published in Mexico in 1935); Octavio Paz, *The Labyrinth of Solitude and the Other Mexico, Return to the Labyrinth of Solitude, Mexico and the United States, The Philanthropic Ogre*, Grove Press, New York, 1985; R. Diaz-Guerrero, *Psychology of the Mexican Culture and Personality*, University of Texas, Austin, 1967; and Alan Riding, *Distant Neighbors*, Alfred A. Knopf, New York, 1985.

but suspect they are nearly the same) were such that an employer could not easily fire an employee (to do so is very costly for the employer). Aware of these costly limitations, *some* employees tend to develop a less than friendly attitude if, for whatever reason, they think you (the customer) have offended them or acted somehow inappropriately. Yet the subject of quality and quality of service is certainly not unknown in France. One can find many books (translated and original works) on the subject of quality or quality of service published by at least four or five French publishers. Apparently, the gap between theory and practice needs to be further reduced.

Still, I must emphasize that the above observations are based on American expectations of what would be considered proper shopping behavior. An important element of the analysis that is totally ignored in the above assessment is the element of shopping culture. French shoppers do not behave exactly the same way as American shoppers. The shopper–clerk interaction is not quite the same as in the United States. The fast-paced Parisian lifestyle—well known to anyone who has driven in Paris—dictates a particular set of behaviors that at times contradicts the more laid-back attitude of many Americans.

Naturally, this does not mean that it is impossible to find pleasant employees in Paris (I have met many). Nor would I want to suggest that there are more unpleasant than pleasant employees in France or that the problem (if it is indeed a problem) is limited to France. That would be erroneous. In the United States, most of the staff in stores are viewed by Europeans as being very friendly (as least when compared to European standards), but not necessarily very helpful and this tends to annoy foreign shoppers. I merely focused on France because I had the pleasure of working there for almost a year; however, similar, if perhaps less blatant, cases of customer neglect could also be made about the United Kingdom where I also lived for a year.

Quality: Absolute or Relative?

Quality professionals often speak of "quality" as if it were an absolute concept and yet quality, whatever its definition, is relative; relative to experience, culture, habits, and so on. What is perceived as being the best in one country may well be viewed as a very inferior product in another. For example, Americans are the only people in the world who think that their coffee tastes good. The rest of the world considers "American coffee" as slightly inferior in taste to cod liver oil. Why? Relativity.

Although there are now numerous gourmet coffee houses and retail stores throughout the United States (especially in Seattle), for decades, most Americans had never experienced the taste of what the rest of the world knew was better coffee. Of course, when we speak of coffee a distinction must be made between different strains of coffee beans as well as different types of roasts. The Robusta strain, which is grown in lowlands and has twice the amount of caffeine as the Arabica strain, is generally considered inferior to the (more expensive) Arabica strain, which is grown at higher elevation. American roast also produces a lighter coffee, which contains more caffeine than the darker roasts (e.g., Espresso, Dark French). Cheap American coffee comes from Robusta Colombian beans. Even today, most of rural America would not trade their cup of Maxwell House for a cup of premium Colombian or Brazilian coffee.[15]

When a Smile Is a Sign of Inferiority

The witty social commentaries offered by Slavenka Drakulic on life after communism cover a broad range of themes. In a chapter entitled "A Smile in Sofia," Drakulic focuses her attention on the relativity of quality of service as defined in the "nonsmiling" culture of employees at the Sheraton Hotel in Sofia, Bulgaria:

> *Here, receptionists, bellboys, lift attendants and waitresses do not smile, not even when you are paying the $260 a night that makes the Sheraton in Sofia probably the most expensive hotel of its kind in the world. Behind their desk in the elegant marble hall with its distinctive Western flavor, the receptionists behave like princesses. Perhaps they almost feel that the hotel belongs to them. After all, until recently, didn't everything here belong to the people? When a guest checks in, they look at him with an air of slight but unmistakable irritation, as if to say, "Who are you? What do you want?" Moreover, they make it clear that you must earn any kindness, as a kind of personal reward.[16]*

Later on in the chapter, Drakulic explains how she was shocked the first time she heard a flight attendant on a PanAm flight bound for New York actually thank the passengers, even adding that she was proud to serve them. How could anyone be proud to serve? Drakulic asks. "In my Eastern vocabulary, it could mean only servitude, slavery, humiliation; something unpleasant and definitely negative."[17] In Sofia, Drakulic explains, a smile is not a sign of courtesy but rather of inferiority. Consequently, although capitalism has invaded Bulgaria and the rest of the former Eastern bloc countries, some of its fundamental principles are not yet understood: "[T]he tenet that a customer is always right, for example, is unheard of."[18]

Quality is not only relative across space but also across time. In an article published in June 1964, author M. L. Katke (of the Ford Motor Company) confidently proclaimed that "our customers have the best."[19] Using a classic biological analogy, Katke spoke of the survival of the fittest and the need to adapt to the needs and desire of the customer. Despite his optimism, Katke perhaps sensed that not all was well at Ford when he prophesied that "The manufacture who does not keep pace with competition in quality as well as price, is in the throes

---

[15]A Brazilian friend of mine once explained why Americans do not know about Brazilian coffee. Many Colombian plantations, he explained, are owned by American companies. Brazilian plantations are owned by Brazilians. Still, I have had Brazilians admit to me that Colombian coffee was better than Brazilian coffee—a clear case of relativity.

[16]Drakulic, *Café Europa*, pp. 46–47.

[17]Ibid., p. 48.

[18]Ibid., p. 50.

[19]M. L. Katke, "Customer Quality Requirements," *Industrial Quality Control*, June 1964, p. 6.

of decadence ... pathway to oblivion."[20] Of course Katke was right because less than 20 years later, the U.S. automotive industry almost went bankrupt thanks to the advances made by innovative Japanese car makers.

Still what is revealing about Katke's comments, which mirrored the jubilant optimism of America during the 1960s, is that he truly believed that "our customers have the best." But did American motorists have the best cars in the 1960s? Certainly, American cars were a desired commodity in many parts of the world. Often, the desire to own an American car reinforced some peculiar status symbol. Still, having been raised overseas during the 1950s and the early 1960s, I recall that one of the most frequent complaints about American cars was that they were too big and the suspension was too soft—which led many people to refer to American cars as "boats." Most European drivers preferred and still prefer a firmer suspension. Despite the criticism, many Europeans liked American cars precisely because they were big, prestigious to own, and comfortable, despite being expensive to operate because of their voracious appetite for heavily taxed gasoline.

All things considered, I doubt that in 1964 everyone in the world (particularly Europeans) would have agreed with Katke's statement. I do not know what most Americans would have thought of Katke's statement, but one thing is certain: They probably did not have all of the necessary information required to reach an informed decision nor did they know that in the 1960s, Detroit did not really care about what they (the customer) desired in a car. As Theodore Levitt observed, in the 1960s Detroit never researched the customer's wants. "It only researched his preference between the kinds of things which it had already decided to offer him, for Detroit is mainly product-oriented, not customer-oriented."[21] Moreover, when I arrived in the United States in 1964, one of the first things I noticed was the almost absolute absence of European cars. The French car maker Renault (then 100 percent owned by the government) had unsuccessfully tried to introduce Americans to its tiny (even by European standards) Dauphine, a car that did not have a good reputation even in France. During the early 1960s, an age that could be characterized as the age of over-abundance (compared to Europe), small cars were certainly not fashionable in America. Given that the American car market in the 1960s was dominated by American cars, one may well ask: Were Fords the best car in the world or were they perceived as being the best car simply because America's economic isolationism and Europe's general inability to (then) produce reliable and comfortable cars prevented American consumers from comparing products?

To complete this story, I should also describe how some Europeans perceive the quality of their cars. When I had the opportunity to work and live in England and France from 1988 to 1990, I noticed that most Frenchmen drove Renaults, Peugeots, or Citroens. In Italy, Fiat dominated the market. In Germany, everyone seemed to drive a Mercedes or a BMW (certainly excellent cars). With the exception of the United Kingdom where Japanese cars are assembled, one

---

[20]Ibid., p. 6.

[21]Levitt, as quoted in Christopher Lorenz, *The Design Dimension*, Basil Blackwell, New York, 1986, p. 28.

notices very few Japanese cars on European freeways (an observation still true in 1998). This is to be expected because most European countries protect their car industries by establishing Japanese car quotas (at least they did in the early 1990s). In France where Japanese cars are a rarity, French drivers are in a position similar to that of Americans in the 1960s: They have little opportunity to compare their Renaults, Peugeots, and Citroens with Toyotas, Nissans, and so on. I have found it interesting to observe that when it comes to automotive "quality awareness" French drivers are generally less demanding than Americans. In other words, features that are considered important in the United States (e.g., quietness, comfort) may not be as important in France or, perhaps more accurately, are rated differently or not necessarily with the same degree of importance.[22] Of course, one must recognize that what the average American considers as essential features of quality may perhaps not be viewed as equally important for the average French driver who is likely to like "nervous" cars, that is, a car with a small yet powerful engine such as the small Peugeots, which are certainly fun to drive and can outperform any small Japanese car. Countless other examples of relativity in quality could be given.

Conclusion

During the 1980s and 1990s, as multinationals have tried to create a global homogeneous market (e.g., world car), they have had to transfer to various parts of the world not only their technologies, management style, and culture but also the quality principles associated with these technologies. A by-product of this technological and ideological transfer has been the attempt, achieved with various degrees of success, to standardize quality methodologies, philosophies, and ideologies. Beginning in 1987, one manifestation of this attempt to globalize quality management principles has been the ISO 9000 series of quality standards, which in a span of 12 years achieved modest, yet significant success (150,000 firms certified worldwide as of 1998).

Quality professionals, some of whom have played an important role in trying to promote the ISO 9000 series of standards, have, to some extent, also played a major role in attempting to standardize an ideology of quality. Unfortunately, because quality professionals favor many ideologies, the process of standardization has been difficult, if not confusing, to the uninitiated who have tried to keep up with the various fads or quality phenomenon.

---

[22]Naturally it is difficult to assess who is right and I am not aware of any survey that has been conducted on this very topic. However, when I visited Argentina I noticed that almost all taxis in Buenos Aires were either Peugeots or Renaults (which I believed are built by AutoLatina in Argentina). I had never seen so many Peugeots or Renaults. Based on this observation, one might think that these are the preferred car in Argentina. Not at all, for when I asked a couple of taxi drivers how they liked their cars they both said that the Peugeot and Renault were cheap cars in constant need of repair and they could not wait for Chrysler to start building cars in Argentina so that they could at last buy a decent car. (Chrysler recently opened a plant in Campo Largo south of Sao Paulo.) Still, one wonders if Peugeot and Renault would blame their Latin American plants for the bad reputation.

Yet what I have attempted to demonstrate in this and other chapters is that, no matter how many books are translated and no matter how uniform or homogeneous the message as to the virtues of quality may be among quality professionals, quality—whatever its definition—is a relative, not an absolute concept, that is influenced by many factors such as culture, experience, multinational corporations, exposure (via multimedia), and economic competition.

One cannot deny that thanks to the ISO 9000 series, concepts of quality and principles of quality have been standardized to a limited extent throughout the world. Yet, although the concepts have been implemented throughout the world, those of us involved with the ISO 9000 series know that the certification process itself is, despite attempts to minimize variation, anything but standardized. Consequently, ISO 9000 certification generates a broad spectrum of emotions ranging from euphoric to sarcastic. Moreover, one simply cannot assume that all requirements of the ISO 9000 standards are understood and implemented equally throughout the world. One need only look at the contractual requirements specified in paragraph 4.3 of ISO 9001-1994 and wonder how a quality manager working in a small to medium-sized company in Algeria, Tunisia, Morocco, Egypt, or any other country of the Middle East would be allowed (by management) to develop such a "contract review" procedure. I do not mean to suggest that the manager would be unable or would not be allowed to develop such a procedure. What I wish to convey is that, in countries where bargaining has been developed to a refined art, where the concept of a fair price is substantially different from Western concepts, and where a handshake or a man's word is worth more than a piece of paper with a signature, the daily application of something known as contract review will likely be different (not necessarily better or worse, but certainly different) from the daily practices of an American or Western European firm. As Richard Pells so eloquently states in his *Not Like Us*, as Europeans have absorbed American culture they also have transformed it to satisfy their cultural needs.[23] Pells comments would apply to all parts of the world and would certainly include aspects of quality that are influenced, modified, and transformed by the culture they invade.

---

[23]Pells, *Not Like Us.*

CHAPTER 4

How Old Can a Company Hope to Be?

I remember, some years ago, conducting a seminar on ISO 9000 in Seattle. As always, I had asked participants to introduce themselves. As we went around the table, one of the participants informed us that his company had been founded in 1894. In all my years of consulting I had never come across a company that was more than 100 years old. Prior to that time, the oldest company I had worked with was a Seattle crane manufacturer founded in 1901. Surveying the other participants, I discovered that none of the other companies was more than 35 years old and a few were barely 10 years old. This rather mundane fact seemed to confirm what I had suspected for some time: that companies, like most living organisms, have an average life expectancy of no more than a few decades. Most "live" to be three, four, or five decades old. Some, mostly small businesses in the service sector, do not survive beyond their fifth birthday, and fewer still live to be 100 or more years.

Why are some companies older (and presumably more successful) than others? Over the years, a few quality gurus and authors have attempted to answer this simple question. Naturally, each author favored his or her own theory for achieving corporate longevity. The paradigm shift theory popular in the 1980s was the key to rejuvenation. During the mid-1990s, reengineering experienced a brief but intense interest on the part of many companies that were anxious to adopt a methodology that would help them justify their downsizing efforts. Total quality management (TQM), which emerged in the 1970s but found its roots in the 1950s with Armand Feigenbaum's principles of total quality control (TQC), continues to be popular to this day. Some consultants would have us believe that a company without a mission statement is bound to fail. This has led the cartoonist Scott Adams to produce some humorous cartoons. Others believe that the primary ingredient to corporate success is strong leadership (hence the myriad of books on leadership). Still others believe that leadership had nothing to do with success.

Could all of these experts be correct? Although each of the opinions and theories just mentioned no doubt has some value, the fact remains that some companies last longer than others not because they believe in paradigm shifts, reengineering, or ISO 9000, but because random events will ensure that a chosen few will last longer than others. (See the next chapter for two specific examples.)[1]

The idea that random events may well be one of the major explanatory variables (but probably not the only variable) for a company's longevity is easier to accept when one refers to the evolution of species. "The history of life," Stephen Jay Gould tells us, "is therefore a story of decimation and limited survival (with enormous success to a few of the victors, insects, for example), not a tale of steady progress and expansion. Moreover, we have no evidence that survivors prevailed for any conventional cause rooted in anatomical superiority or ecological adaptation. We must entertain the strong suspicion that this early decimation worked more as a grand-scale lottery than a race with victory to the swift and powerful. If so, then any rerun of life's tape would yield an entirely different set of survivors."[2]

What is interesting about Gould's comments is his reference to a "grand-scale lottery," with enormous success to a few of the victors (e.g., Microsoft, Boeing, GM, General Electric, Sony, etc.). Applying Gould's observation to the industrial evolutionary history of the United States, one must conclude that if the history of the last 200 years could be replayed, there is no guarantee that the same "corporate winners" would emerge. A similar conclusion is reached by James C. Collins and Jerry I. Porras in their popular *Built to Last: Successful Habits of Visionary Companies*. However, rather than simply state that the longevity and thus long-term success of certain companies are primarily associated with luck, Collins and Porras explain longevity, in part, by relying on the artificially contrived concept of visionary companies. Visionary companies, we are told, are "premier institutions—the crown jewels—in their industries, widely admired by their peers and having a long track record of making a significant impact on the world around them."[3] The dubious use of the word *visionary* seems unnecessary, particularly when a few pages later the authors tell us that "Visionary companies make some of their best moves by experimentation, trial and error, opportunism, and—quite literally—accident. What looks *in retrospect* like brilliant foresight and preplanning was often the result of 'Let's just try a lot of stuff and keep what works.' "[4] One is left with the definite impression that so-called "visionary com-

---

[1]The concept of randomness and its effect on financial investments is explained in Burton G. Malkiel, *A Random Walk Down Wall Street*, W. W. Norton & Company, New York, 1990. Malkiel, who has little faith in financial managers, suggests that a naive buy-and-hold strategy using a dart-board-selected portfolio can provide a 10 percent return on your investment (p. 137). See also Frank Partnoy, *Fiasco: The Inside Story of a Wall Street Trader*, Penguin Books, New York, 1999.

[2]Stephen Jay Gould, *Eight Little Piggies. Reflections in Natural History*, W. W. Norton & Company, New York, 1993, p. 225. It is interesting to note that James Collins and Jerry Porras (authors of *Built to Last*, HarperBusiness, New York, 1994) are also influenced by Jay Gould but reach slightly different conclusions. The French biologist Jacques Monod offered in the late 1960s a similar theory regarding evolution; see his *Le Hasard et la Necessite*, published in English as *Chance and Necessity: An Essay on Natural Philosophy and Biology*, Vintage Press, New York, 1971.

[3]James C. Collins and Jerry I. Porras, *Built to Last: Successful Habits of Visionary Companies*, HarperBusiness, New York, 1994, p. 1.

[4]Ibid., p. 8. The concept of luck is developed further by the authors under the section titled "Corporations as Evolving Species," pp. 143–150.

panies" are more willing to experiment and are perhaps luckier than other companies, which is why they have endured longer than their rivals (and not necessarily because they have any vision whatsoever). Moreover one wonders what Collins and Porras' list of visionary companies would look like had they conducted their studies in the 1960s. Had they had done so they would have probably agreed with M. L. Katke (of Ford) who in the June 1964 issue of *Industrial Quality Control* stated that "there never was a truer application of the theory, 'Survival of the fittest,' than in the automotive industry."[5] With some guarded confidence Katke went on to proclaim that "At the moment, the quality criterion of the American people is being met in our industry."[6] And yet, perhaps sensing that all was not well within the automotive industry, Katke warned with great foresight that "The manufacturer who does not keep pace with competition in quality as well as price, is in the throes of decadence . . . pathway to oblivion."[7] Fifteen years later, the Ford Motor Company nearly experienced oblivion.

Still, Collins and Porras do provide their readers with some valuable contrasting observations between what they call visionary companies and other not-so-visionary companies. One of the most significant points made by the authors and recognized by them as "one of the most fascinating and important conclusions" was that "creating and building a visionary company absolutely does not require either a great idea or a great and charismatic leader."[8] Flexibility, a willingness and ability to experiment and try many new ideas, the wisdom to recognize and discard bad ideas and/or experiments and adopt good ideas, and an element of random good luck seem to comprise the basic elements of so-called visionary companies.

I would partly agree with Collins and Porras when they suggest that luck is an important element of corporate longevity. However, unlike them, I do not believe luck applies only to visionary companies: Luck has no preference, it favors no one. Indeed, many of the very successful companies found in California's Silicon Valley—many of which are barely 3 or 4 years old—have achieved extraordinary growth thanks to their technological ideas and innovations rather than a vision.

Still all of this analysis does not tell us anything about the average life expectancy of companies. The answer to that question is not easily obtained. Statistics on when a company was founded are found in various publications (*Standard and Poors*, for example). I was able to discover a company profile database that contains 175,000 companies and includes, among other things, the date on which a company was founded. I performed two searches, one for the general category of manufacturing and one for the general category of chemical (see Tables 4–1, 4–2, and 4–3).

---

[5]M. L. Katke, "Customer Quality Requirements," *Industrial Quality Control*, June 1964, p. 4. Collins and Porras, *Built to Last*, classify Ford as a visionary company and GM as a nonvisionary company. That is probably a correct assessment but GM, despite many setbacks, still dominates the automotive industry.

[6]Katke, "Customer Quality Requirements," p. 5.

[7]Ibid., p. 5.

[8]Collins and Porras, *Built to Last*, p. 25.

TABLE 4–1 Manufacturing Companies Founded by Decade (1770–1997)

Decade	Number of Companies Founded during the Decade	Percentage	Cumulative Percent
1770–1780	4	0.012	
1781–1790	9	0.026	0.038
1791–1800	12	0.035	0.073
1801–1810	19	0.056	0.129
1811–1820	23	0.06	0.189
1821–1830	33	0.09	0.279
1831–1840	74	0.22	0.499
1841–1850	138	0.41	0.909
1851–1860	179	0.53	1.439
1861–1870	269	0.80	2.239
1871–1880	417	1.24	3.639
1881–1890	687	2.05	5.689
1891–1900	796	2.38	8.069
1901–1910	1,262	3.77	11.839
1911–1920	1,534	4.59	16.429
1921–1930	1,964	5.87	22.299
1931–1940	2,119	6.34	28.639
1941–1950	3,648	10.92	39.559
1951–1960	3,997	11.96	51.519
1961–1970	4,938	14.78	66.299
1971–1980	4,668	13.97	80.269
1981–1990	5,465	16.35	96.619
1990–1997	1,146	3.43	100.0
Total 1770–1997	33,405		

Source: Company profiles database of 175,000 companies.

TABLE 4–2 Percent of Manufacturing Companies that Were 100 Years Old by Decade (from 1990–1940)

Time Period	Number of Companies	Estimated Percent of Companies that Were 100 Years Old
1770–1990	32,259	In 1990s 2.46%
1770–1980	26,794	In 1980s 1.55%
1770–1970	22,126	In 1970s 1.21%
1770–1960	17,188	In 1960s 1.04%
1770–1950	13,191	In 1950s 1.04%
1770–1940	9,543	In 1940s 0.07%

Source: Company profiles database of 175,000 companies.

TABLE 4–3 Chemical Companies Founded by Decade
(1770–1997)

Decade	Number of Chemical Companies Founded during the Decades
1770–1780	0
1780–1790	0
1790–1800	0
1800–1810	0
1810–1820	1
1820–1830	1
1830–1840	2
1840–1850	5
1850–1860	6
1860–1870	4
1870–1880	11
1880–1890	15 (or 1.6% of 918)
1890–1900	24 (or 2.6% of 918)
1900–1910	28
1910–1920	50
1920–1930	65
1930–1940	84
1940–1950	103
1950–1960	117
1960–1970	140
1970–1980	123
1980–1990	125
1990–1997	20
Total 1810–1997	918

Source: Company profiles database of 175,000 companies.

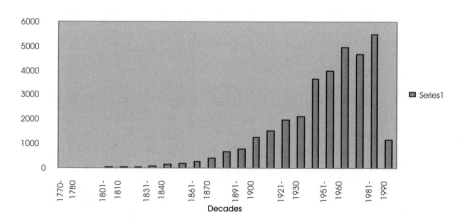

FIGURE 4–1 Manufacturing Companies Founded by Decade (1770–1997)

Reviewing the data collected in Table 4–1 and Figure 4–1, one notices that only a little more than 2 percent of the sample of 33,405 manufacturing companies are 100 years old (a similar percentage is found in Table 4–3 for the chemical industry). Moreover, as Table 4–2 reveals, the percentage of 100-year-old firms varies by decades between 1 and 2 percent. Although the percentage does seem to have increased over the decades, it is too early to tell if a statistically significant trend is present.

What the statistics presented in Tables 4–1 and 4–3 reveal is that, at least within the United States, very few companies (within the manufacturing and chemical sectors) are likely to survive to their 100th birthday. Are these companies visionary companies? I doubt it, but to find out would require several months of investigation. Indeed, for the 1891–1900 period alone, 796 companies would have to be analyzed. If we include other decades (1940s to 1990s), the number of companies that would have to be analyzed approaches 2000. Not an impossible task for a team of graduate students; unfortunately, I do not have the luxury of such assistance.

Nevertheless, based on what I recall seeing flashing on the computer screen, I would hypothesize that no unique pattern of "visionary" actions is likely to be found in these 100+-year-old companies. I formulate this hypothesis based on the recollection that several of the companies I saw were in particular types of industries such as extracting or metal processing, for example—industries not generally known for any particular form of "visionary" activity.

If one were to further investigate my tentative hypothesis, one would first have to collect all the names of the companies that are 100 years old as of 1990 (for example) and try to classify the companies by type of industry (or perhaps standard industrial codes). The process could be repeated for the 1980s, 1970s, 1960s, and so on, to see if the same pattern of classification recurs. Once the classification is complete, the next task would be to see if a certain type of industry is more likely to outlive other industries—I would propose that the answer to that hypothesis is "yes."

CHAPTER 5

Built to Last for a While: The Age of Flexibility

We first sell the drives; then we design them; and then we build them.
Executive of Conner Peripherals quoted in Clayton Christensen's
The Innovator's Dilemma[1]

Whenever I visit the San Fernando Valley in Southern California, which is at least once a year during Christmas, I am always amazed at how rapidly the economic landscape changes from year to year (not to mention the physical landscape, which is rearranged periodically by earthquakes). Property values, which had fallen by as much as 35 percent during the 1992–1993 recession, were once again going up at an alarming rate (1997–1998). The recession is nothing but a faint memory awaiting to be awakened by the Asian financial and economic crisis that began in October 1997. Where there used to be a gas station, there are now half a dozen stores; old apartments have been torn down and replaced by condominiums. Whole cities are still mushrooming inland from San Diego, despite the 1992–1993 recession, which hit California harder than most other states. The continuous recycling and partitioning of the Californian economic landscape, and the resulting increase in value, seem unending. Once again the economic bubble is expanding.

Continuous and rapid change seems to be the constant in California. This ability to thrive on turbulence was noted by the British weekly, *The Economist*, which commented that California is the state with the most "gazelle" firms. Gazelles, in the management jargon of the mid-1990s, are firms that can double their size in 4 years. These firms have a median age of 15 years, are responsible for 70 percent of America's job growth (but are no guarantee for stable employment), and "they are inherently experimental creatures, exploring new markets with new products, succeeding sometimes, messing up frequently, recovering—and starting the process all over again."[2] *The Economist* concluded by noting that

---

[1] Clayton Christensen, *The Innovator's Dilemma: When Technology Causes Great Firms to Fail*, Harvard Business School Press, Boston, 1997, p. 138.

[2] "Gazelle Firms," *The Economist*, May 28, 1994, p. 65.

America's most dynamic regions (California, Nevada, Colorado, Arizona, and Florida) are all states with risky corporate survival but with fast track.

Flexibility and Competitiveness

If we are to believe the many preachers of quality, quality cannot be achieved unless you implement the following series of actions (not necessarily in this sequence): You must empower your employees, teach them to respect or even adore customers, and emphasize the value of just-in-time delivery. As they begin to apply various statistical techniques to continuously improve processes, employees should also be trained in quality function deployment. This will allow them to better listen to the so-called voice of the customer—a voice that is rarely, if ever, a uniformed chorus as consultants would have us believe but, rather, a cacophony of varied and at times contradictory wishes. Still, this cannot be achieved unless employees reorganize themselves into quality action teams (not quality circles), whose purpose would be to constantly innovate and rediscover themselves. This cannot be accomplished unless employees first analyze and understand the psychological profile of their coworkers. Armed with such valuable information, they can now motivate each other. As they begin to reengineer their firm, they should simultaneously try to benchmark other organizations and thus absorb the best practices (relevant and irrelevant) of others. Finally, executive management should really learn to become better leaders, better yet, they should become *change agents*.

But, just in case all of these actions should be inadequate, companies should really register their firm to one of the ISO 9000 standards—or QS 9000 standards for suppliers to the automotive industry, win the Malcolm Baldrige Award, win their state quality award (now offered by an increasing number of states), and always be on the lookout for more awards (the national Quality Cup offered by *USA Today* might not be a bad idea).

However, these commitments might not be enough to appease the demands of environmentalists and state regulatory agencies. Consequently, to project a positive image, companies should also register their environmental management system to the ISO 14001 standard. Of course, this does not mean that they will no longer have to deal with state and federal environmental regulators, but it does ensure that companies will now have to welcome more auditors who hopefully will not contradict each other.

Among this plethora of management systems a firm must also learn to accurately define and manage its internal costs. Because the management of all of these systems is likely to become a difficult proposition, the use of sophisticated enterprise-wide and fully integrated information software systems such as SAP, Oracle, and others is now a (costly) must. Naturally, more training will be required for any company entering the costly world of enterprise resource planning (ERP). It is a wonder companies find the time to make products and earn a profit. No wonder the average workday is increasing.

If this sounds ridiculous, don't laugh, for it might well describe the various programs undertaken at your company during the late 1990s. This need, by

American managers, to constantly be in search of the quick fix has been very eloquently described by Michael McGill in his *American Business and the Quick Fix*. "Managers," McGill concludes, "must give learning the same priority as doing. The tendency of managers is to get caught up in doing, doing, doing, never pausing to process what they are doing so that they might learn from their experience."[3]

But is there any evidence to suggest that implementing the types of quick fixes mentioned will truly improve a company's overall long-term performance? Certainly, we have seen examples of short-term success stories, but clearly one must also acknowledge that the business section of the daily press is full of stories about "quality organizations" which, because of a multitude of known and/or unknown causes, have had to lay off hundreds and sometimes thousands of workers. Surely, the reasons for success cannot be solely attributed to so-called total quality management practices as preached by quality professionals. Other factors must be considered.

Economic Success of the Firm: Is It Based Solely on Quality Issues?

Mark Minasi, author of *The Software Conspiracy*, would undoubtedly answer that the economic success of a firm, or even a whole industry, is certainly not dependent on the quality of its products.[4] The $25 billion U.S. software industry, not known for producing defect-free software, continues to attract, year after year, millions of consumers in search of the latest software upgrades complete with the "usual" number of defects known as "bugs." What is astonishing is that many programmers and company executives within the software industry claim that customers are not really interested in "bug-free" software; rather, they would claim that customers want to buy the latest upgrades. What is even more astonishing is that millions of customers have come to accept the meaningless error messages by routinely clicking on the "OK" button that often appears on their screen. Yet, the software industry, where the basic concepts of quality control or process measurements are apparently either not known or generally ignored by software engineers, continues to thrive on Wall Street.

Competitive advantage, Michael Porter contends, is attributable to highly localized factors characterized by differences in national economic structures, values, cultures, institutions, and histories. After analyzing the economies of several countries, Porter observed that successful firms tend to be concentrated in particular cities or states within a nation. To be successful, firms must establish successful linkages with their suppliers and distributors. Thus, a firm having an excellent product but a poor distribution system is not likely to survive very long. In many cases, firms gained a prominent position simply because of geographic circumstances. Swedish dominance in prefab houses was partly a result of Sweden's rigorous winters. Brazilian road conditions are so demanding that Brazilian manufacturers of shock absorbers have been able to establish them-

---

[3]Michael McGill, *American Business and the Quick Fix*, McGraw-Hill, New York, 1988, p. 219.

[4]Mark Minasi, *The Software Conspiracy*, McGraw-Hill Book Company, New York, 1999.

selves as world leaders in the manufacturing of durable shock absorbers. Government regulations can also favor or retard innovations. In the United States where the manufacturing of plastic gas tanks is forbidden by law, Americans are forced to drive cars whose tanks are less safe than in other countries. Indeed, plastic tanks have proven to be more reliable and significantly sturdier than their metal counterpart. In other cases, a symbiotic relationship can help a particular company and product. Earth-moving equipment, for example, benefited from the worldwide activities of American construction firms and mining companies. The training of foreign doctors and dentists in the United States has helped promote the purchase of U.S. medical and dental equipment overseas. Finally one could cite the sale of computers overseas, which simultaneously helped manufacturers of computer peripherals and software. In such cases, however, the advantages usually last only 2 to 3 years until local manufacturers begin to compete.[5]

Porter's analysis focuses on businesses that operate in a competitive market. But what about businesses that operate in a new, noncompetitive market? Do they obey the same laws? One example of a business that operates in a still marginally competitive market is the Internet service provider (ISP). Over the course of 4 years in the late 1990s I had four different Internet service providers. I once had a communication problem with my ISP. Every once in a while, I would get disconnected after a few minutes. After almost 5 months of complaining about the problem, I was getting nowhere. The ISP insisted that everything was fine at their end and they could not help me. Perhaps it was my modem or my software or maybe I should restart the computer. Then, one day, the ISP installed new communication equipment and, miraculously, all of my problems went away! No more modem problems.

It has been my experience that although service has since improved, ISPs as well as the emerging multibillion dollar electronic business (e-business) industry found on the Internet, can be characterized by certain common features:

1. They do not really understand the meaning of customer service. They all have technical support groups that are staffed with friendly people but, for the most part, they have either a limited knowledge of communication protocol issues or they simply are not willing to admit or are told "never to admit" that perhaps the system is slow because the company oversubscribed its resources.
2. The general assumption of all ISPs I have had to do business with is that, if there is a problem, it is not at their end but rather at "your end." Their modems and systems are invariably working perfectly. Thus, if there is a problem it is likely to be with the phone company or more likely "your computer." Perhaps you should check this or that, reverify some settings that you never modified in the first place, and so on. One should note that a similar philosophy of "customer service" is found within the software industry. Despite poor customer relations, ISPs are still thriving.

---

[5]Michael Porter, *The Competitive Advantage of Nations*, The Free Press, New York, 1990, pp. 19, 42–43, 71, 84, 98, 105.

Quality: One Small Element to Economic Viability

During a period of nearly 40 years, from the early 1900s to the late 1940s, the economist Joseph Schumpeter published a series of influential books on business cycles and the workings of capitalism in general. By reformulating and sometimes refuting Karl Marx's arguments regarding the eventual demise of capitalism, Schumpeter's historical analysis led him to believe that the capitalist engine was kept in motion via new methods of production, new markets, new forms of industries, organization, and innovations (which he saw increasingly as being under the control of large corporations). Schumpeter also believed that one of the normal consequences of capitalism was a process that he called "creative destruction" by which businesses are constantly destroyed to create new ones.[6] Schumpeter's insight into the workings of capitalism are difficult to refute when one observes the recent wave of mergers and/or acquisitions, which invariably leads to the creation of new organizations and the simultaneous destruction of others.

The process of creative destruction is clearly evident within the high-tech world, where thousands of firms are created and destroyed every year (the same could be said about e-business). Indeed, within the highly competitive economy of the Silicon Valley in northern California, the process of creative destruction can only take 2 to 3 years before a firm is "absorbed" by a larger competitor. In many cases, the very reason for creating a company is to raise enough venture capital to create the company, attempt to develop a product (to support a concept or idea), issue paper shares, and finally hope to eventually be acquired by a larger competitor, thus transforming the founders and their associates into instant millionaires. Numerous cases could be cited in the Silicon Valley surrounding San Jose, California. Often, the success of the company is never really measured in terms of quality of the product; in fact, the product may even be of poor quality but have great potential! In some cases (Internet and e-commerce), the acquired company may not even produce a product but only an idea!

Schumpeter's theory complements the theories of natural evolution of Monod and Gould, discussed in an earlier chapter, which propose that the probability of success of any organism can be caused by nothing other than chance and necessity. Similarly, exogenous economic factors can randomly destroy efficient and inefficient firms. October 1997 was not a very good month for many Asian firms. As the stock markets in Bangkok, Seoul, and Hong Kong began to decline rapidly—almost overnight—a domino effect took place as other Asian markets in Singapore, Indonesia, and even Japan. Many companies were ruined overnight in Korea, Thailand, and other Asian cities. Did these companies fail because their products were of poor quality? Did they fail because they were not ISO 9000 registered? Or did they fail because management was not committed to one or more of the quality paradigms preached by so many experts over the years? Probably not. Many companies went bankrupt and had to lay off thousands of employees just to survive because financial and political economic circumstances outside their control forced them to do so. Poor leadership was not necessarily the cause

---

[6]Joseph A. Schumpeter, *Capitalism, Socialism and Democracy*, 3rd ed. Harper and Row, New York, 1947, pp. 81–106.

for their downfall. Moreover, and contrary to what quality consultants and others have claimed for many years, the so-called "leveled field" that was supposed to be achieved thanks to ISO 9000 certification did not immunize these companies to bankruptcy. When regional or global economic forces take over, there is little that a company (old and young, certified or not) can do to avert disaster.

The fragility of the network of financial dependencies and economic interdependencies woven by multinational corporations as they helped bring about globalization soon became apparent. Within 6 to 8 months the effects of the Asian financial and economic crisis was being felt in the United States. In California and other western states, newspapers began to write about the impact of the Asian crisis on the local economies (a crisis that a few months earlier experts had predicted would not happen). Here is a sample of the headlines:

- "American Factories Suffering Fallout from Asia's Financial Woes," *The Press-Enterprise*, June 25, 1998, p. E4.
- "Inland Purchasing Managers Expect Economy to Slow," *The Press-Enterprise*, July 2, 1998, p. C8.
- "Rockwell May Cut 10 Percent of Its Workforce: Asia's Financial Crisis, Drop in Semiconductor Business Cited," *The Press-Enterprise*, June 27, 1998, p. H3.
- "Motorola: Bad to Worse? Struggling Chipmaker to Post Rare Loss," *USA Today*, July 7, 1998, p. B1. In this article, the Asian crisis was cited as one of the causes for the elimination of 15,000 jobs.

Poor quality of product had little if anything to do with the demise of many Asian firms. Poor financial decisions and bad loans spread over many years were more likely the cause. And yet, even if a company has made sound financial decisions over the years, it is not immune to possible bankruptcy. In the United States, under the personal injury system of law (also known as tort law), financially successful companies that abide by all of the latest tenets of management quality can be bankrupted in a matter of a few months.[7] Unlike what most quality consultant experts would have us believe, better quality often cannot help a company because product quality is not always the principal reason for economic failure. Let's look at an example of two companies (one 65 years old and the other 120 years old) that have experienced very different success stories.

G. H. Bass versus Vita Needle Company

The newspaper heading immediately caught my attention: "An American original has lost its footing."[8] The article described how the well-known shoe manufacturer, founded more than 120 years ago by George Henry Bass, was shutting down its Wilton, Maine, plant. The reason given for the shutdown was classic:

---

[7]Max Boot, *Out of Order*, Basic Books, New York, 1999; see particularly Chap. 6, "The Civil Injustice System."

[8]Fred Bayles, "An American Original Has Lost Its Footing," *USA Today*, February 10, 1998, p. 4A.

international trade and decades of competition from low-wage, offshore factories. About 350 workers in this town of 3900 were to lose jobs that had been in their families for generations. Ironically, G. H. Bass was moving its operations to Puerto Rico and the Dominican Republic, the very offshore islands that had brought about its demise.

Wanting to know more about G. H. Bass & Co., I searched the Internet for some additional information. What I found on the G. H. Bass web page—which showed a copyright of 1997—was revealing. Under the "Our Mission" page, one could read the following: "If you're interested in joining the Bass team, take a look at what we have to offer."[9] Checking the "Job Opportunity" subpage, I learned that sometime in 1997, or at most a year before the *USA Today* article appeared, G. H. Bass & Company was looking for a developer, a JIT manufacturing manager, a marketing manager for event planning and promotions, a quality assurance manager for lasting and fit, and a vice president of marketing. Apparently, a few key individuals had already learned that the end was near and had decided to leave the company before the official announcement.

A little over 100 miles south of Wilton the 65-year-old Vita Needle Company located in Needham, Massachusetts, is still turning a profit of $3 million a year. The company, whose workforce averages 73 years of age, originally focused on reusable hypodermic needles. But with the advent of AIDS, demand for reusable needles plummeted and diversification was essential for survival. The company (which is not ISO 9000 certified) began manufacturing state-of-the-art hollow pins, embroidery needles, tips for dart guns, needles to blow up balls, glue applicators, tubes for particle research in nuclear physics, and pressure gauges for aircraft tires. They did all of this with the same well-maintained equipment that has not been updated since 1939! Workers in the metal press average more than 80 years of age. What is unique about the company is that it is still in business. The 72-year-old chairman still gets the mail and handles customer complaints, which are few. "Paperwork is kept in accordion files on a crooked shelf in the common office shared by the president, the chairman and a few other executives. The same wooden roll-top desks, white stucco walls, ceiling fans and frosted glass light fixtures have been there since the 1930s."[10]

Both Bass and Vita Needle made good-quality products, yet one failed and the other continues to be successful. Other failures are listed below:

- In 1892 F. W. Woolworth wrote to one of his store managers: "We must have cheap help or we cannot sell cheap goods" (cited in Harry Braverman's *Labor and Monopoly Capital*, Monthly Review Press, New York, 1974, p. 371). This was no doubt good advice and it worked for 117 years but in July 1997, Woolworth announced that it was closing its 400 remaining five-and-dime stores across the United States and laying off 9200 employees. (The only remaining stores would be in Germany and Mexico.) Faced with com-

---

[9]G. H. Bass web site, www.ghbass.com.

[10]Jon Marcus, "Old-Fashioned Firm Still Turns a Profit," *The Press-Enterprise*, December 30, 1997, pp. C1, C9.

petition from Wal-Mart and other discount stores that began to operate from the suburbs and who had apparently improved on Mr. Woolworth's advice, Woolworth could no longer attract customers to its downtown stores. In recent years retailers have increasingly discovered that customers are always in search of cheaper and cheaper products. Unfortunately for the employees working for these retail shops, cheap products mean minimum wages or, worse yet, job cuts. (See, for example, "Bargain Prices Have a Price: Job Cuts," *USA Today*, August 31, 1993, B1).

- Ernst Home Center, founded in Seattle in 1893, filed for bankruptcy in 1996. Reasons: squeezed by competitors such as Home Depot, Eagle Hardware (*The Seattle Times*, November 23, 1996, p. C1).
- Dydee Diaper Service, founded in 1933, folded in 1996. Reason: Cloth diapers are no longer favored. Environmentally unaware yuppies prefer plastic disposable diapers (*The Press-Enterprise*, December 20, 1996, p. D5).
- A 142-year-old sawmill shut down in 1995. Reasons: Management cited soaring log prices, tough environmental regulations, and competition from Canadian and Japanese buyers who outbid Americans for top-quality logs (*The Oregonian*, August 28, 1995, pp. A1, A8).
- Baring Brothers and Co., a 233-year-old investment bank, folded in 1995. Reason: A 28-year-old "rogue trader," who concealed what he was doing from management, lost more than $795 million (*Seattle Post-Intelligencer*, February 27, 1997, p. 1).

The Dilemma of Responsiveness

In his excellent book *The Innovator's Dilemma*, Clayton Christensen provides a detailed analysis as to why so-called "great firms" can experience major setbacks and occasional failures. Reviewing the case history of several successful (or once successful) companies in the disk drive, excavator, steel mill, and other industries, Christensen summarizes the innovator's dilemma by observing that "Blindly following the maxim that good managers should keep close to their customers can sometimes be a fatal mistake."[11] Indeed, as Christensen illustrates, successful firms can lose their positions of leadership "[p]recisely *because* these firms listened to their customers, invested aggressively in new technologies that would provide their customers more and better products of the sort they wanted, and because they studied market trends and systematically allocated investment capital to innovations that promised the best returns."[12]

---

[11]Christensen, *The Innovator's Dilemma*, p. 4.

[12]Ibid., p. xii. Christensen contrasts disruptive technologies with sustainable technologies: "What all sustaining technologies have in common is that they improve the performance of established products, along the dimensions of performance that mainstream customers in major markets have historically valued" (p. xv). Some of the characteristics of disruptive technologies are that (1) they result in *worse* product performance; (2) they are "technologically straightforward, consisting of off-the-shelf components put together in a product architecture that was often simpler than prior approaches" (p. 15); (3) they are initially embraced by the least profitable customers in the market; and (4) they focus on markets that are too small for the industry leaders who, because of their large market share, are constantly looking for rapid short-term gain in their well-established market.

By trying to continuously please their best and most demanding customers, firms eventually end up developing products whose functionality exceeds the needs of most (emerging) customers. It is at this stage when "performance over-supply opens the door for simpler, less expensive, and more convenient—and almost always disruptive—technologies."[13] (Successful firms often become exposed to the "disruptive technologies" introduced by smaller firms not bound by conventional total quality management thinking.) Christensen's observations are revealing because they challenge the quality paradigms presented over the years. What Christensen suggests is that there are times when doing all the right things as prescribed by TQM is no longer good enough and may even be counterproductive!

Well-established firms are, with few exceptions, unable to develop disruptive technologies because their management structure and philosophy are too entrenched in satisfying sustainable technologies demanded by their well-established and often powerful customers. These firms do not know how to develop disruptive technologies because there is no reliable source of customer input, no existing market. Because most marketers working for these "great firms" have little practical training "in how to discover markets that do not yet exist," it is difficult for these firms to design products (in a vacuum).[14]

Expanding on Christensen's observations, I would also suggest that the strict adherence to a quality assurance system such as ISO 9001, for example, which is designed to manage sustainable technologies, will *prevent* a firm from developing disruptive technology. Christensen suggests that the best way for a well-established and successful firm to successfully develop disruptive technology is to set up an independent (preferably small) organization. This independent organization would be shielded from the needs of well-established customers and would therefore be allowed to focus on the nascent market defined by emerging customers—the very customers who do need the technology "bells and whistles" demanded by the well-established customers.

Conclusion

During the 1990s I visited numerous *successful* companies that were considering applying for ISO 9000 registration. One of the reasons these companies were only considering ISO 9000 compliance as opposed to wanting to achieve third-party registration was that they knew, or at least hoped they knew, that if they waited or rather stalled their clients long enough, the customers would eventually "forget" about their request (a similar, if less successful strategy is currently applied by some suppliers to the automotive industries). Many of these companies waited as long as they could to implement ISO 9000 or QS 9000 because they knew that their current quality assurance system did not satisfy the requirements specified in these standards and, moreover, many questioned the validity, logic, cost, and value of the heavy emphasis placed on documentation.

---

[13]Ibid., p. 195.

[14]Ibid., pp. 147, 195.

Most firms had procedures and records, but the fact that these procedures were outdated apparently did not affect their business—or at least so they thought. Some firms were actually recognized leaders in their industry. A small manufacturer of scientific instruments in Ohio, for example, which designed specialized instruments by practicing concurrent engineering, did not satisfy any of the design control requirements stated within the ISO 9001 model. Yet, although the quality and production managers were very frustrated, the engineering staff always presented them with counterarguments as to why the current system was adequate for their needs. I do not know if the firm in question ever bothered to achieve registration.

Another incident relates to a conversation I once had with a process consultant for medical device manufacturers. During the conversation, the consultant explained that he once visited a firm that designed products for the medical industry. Upon inspecting a prototype the manufacturer had just completed, the consultant asked for the engineering drawings. To his surprise, he was told that there were none. "How did you build the prototype?" the consultant asked. "We went to the machine shop and built it," was the answer. "But you can't do that," the consultant replied, "the FDA won't allow it." The consultant concluded his story by telling me how, by using some well-known software, he was able to reverse engineer the part. But, I ask the reader, what is wrong with reverse engineering? Why do we assume that parts and prototypes must be designed and developed the way some standards arbitrarily prescribe? Does it matter how the part is designed and developed as long as it satisfies the *final* requirements. Surely, when the shop produced the prototype it had to design it in such a way as to simultaneously satisfy a set of (occasionally poorly defined) external customer requirements and the associated internal (engineering) requirements. Thus, even though the set of customer requirements might have been vague or even not defined (as is often the case), a set of guidelines, perhaps a methodology, was followed. But, is it reasonable to suggest that the methodology should be prescribed *a priori* (as QS 9000 and to a lesser extent ISO 9001-1994 and the 2000 versions specify) by the client, who may or may not even have a well-defined set of specifications?

Companies that have learned to specialize for some small market niches have also become successful. Italian firms, 99 percent of which are family businesses, known for their flexible mass production (flexible specialization), have had great success in recent years applying the principles of mass customization, rediscovered by Stanley Davis and others. But even in Italy, some small firms have had to pursue ISO 9000 registration and despite that, it is not certain whether they will be able to survive unless they learn to grow bigger by taking over other businesses and carefully selecting their customers.

CHAPTER 6

On Servicing the Customer

Who Is the Customer?

On a flight from Reno to San Diego, I once overheard a conversation where the flight attendant was telling a passenger (perhaps a friend) that airlines were waiting for a ruling from the FAA (Federal Aviation Administration) that would ban all carry-ons. The flight attendant explained that a few days ago a Continental flight had suddenly dropped about 500 feet because of air turbulence. As a result of the sudden drop in elevation, the door of an overhead compartment had opened and a bag had fallen on a lady and broken her arm. The flight attendant concluded her story by suggesting that obviously carry-on bags had to be banished. To my surprise, the passenger, who had claimed moments earlier that she was a frequent flier, agreed. "Everyone should be required to check their bags, I do," she said. Noticing that the lady looked at me as she voiced her opinion, I decided to interject and explained that I disagreed because I always use carry-ons and do not like to check in my carry-on (mostly because airlines are so slow at delivering bags and suitcases). The flight attendant jumped in to support her friend's argument. I asked who her customers were: passengers or the FAA? Almost without hesitation she replied, "The FAA." I guess the attendant thought that it is thanks to the FAA that she has a job.

The above anecdote would suggest that one of the reasons why some airlines do not want passengers to have carry-ons is because they are afraid of lawsuits. Airlines are probably pressuring the FAA to pass the new rule without concern for passenger convenience. But would it make sense to introduce new regulations without conducting a survey? I would think that the FAA would first like to know how many passengers have had arms broken because of a bag falling on them during air turbulence? Other important questions to ask would include these: How many passengers would like to have the opportunity to carry on small baggage aboard a plane? What is the ratio of these two numbers (people with broken arms versus people who approve of carry-ons)? Should laws be written to protect the small minority of those who might suffer a broken arm? Those who wish to check their bags or suitcases can always do so. I would like airlines to keep the option of bringing carry-on bags on board.

The Case of ISO 9001-2000

A document (ISO/TC 176/SC 2/N 415) dated July 30, 1998, reads as follows (italics added):

1.3 Reasons for a more thorough revision of the ISO 9000 standards

Customer needs are the force driving the revision of these standards. In 1997, ISO/TC 176 conducted a large global survey of 1120 users and customers to better understand their needs. This was accomplished using a questionnaire covering:

attitudes towards the existing standards
requirements for the revised standards
the relationship of the quality management system standards to the environmental management system standards.

The following significant user and *customer needs* were determined from the analysis of these questionnaires:

The revised standards should have increased compatibility with the ISO 14000 series of Environmental Management System Standards.
The revised standards should have a common structure based on a Process model.
Provision should be made for the tailoring of ISO 9001 requirements to omit requirements that do not apply to an organization.
ISO 9001 requirements should include demonstration of continuous improvement and prevention of non-conformity.
ISO 9001 should address effectiveness while ISO 9004 should address both efficiency and effectiveness.
ISO 9004 should help achieve benefits for all interested parties, i.e., customers, owners, employees, suppliers, and society.
The revised standards should be simple to use, easy to understand, and use clear language and terminology.
The revised standards should facilitate self-evaluation.
The revised standards should be suitable for all sizes of organizations, operating in any economic or industrial sector, and the manufacturing orientation of the current standards should be removed.

To ensure that the revised standards satisfy these user and customer needs, a validation process has been implemented. The validation process allows for direct feedback from users and *customers* at key milestones during the revision process to determine how well these needs are being met and to identify opportunities for improvement.

When I first read the above paragraphs, I was amused by the use of the words *customer needs*. Who were these phantom customers who demanded that "[T]he revised standards should have increased compatibility with the ISO 14000 series of Environmental Management System Standards," or that "[T]he revised standards should have a common structure based on a Process model," or better yet that "ISO 9004 should help achieve benefits for all interested parties, i.e., customers, owners, employees, suppliers, and society"? Which rational customer, I wondered, would make such demands? None of my customers knew what the "Process model" was and most did not care about ISO 14000! And they certainly had never heard of any changes nor did they wish any changes to be made to the existing standards. So, who were these 1120 concerned customers who were making these demands?

The results of the survey (available as of late 1998 at www.bsi.org.uk/iso-tc176-sc2/) reveal some interesting facts:

1. More than a third of the respondents (34 percent) are from the United Kingdom. This is to be expected since the United Kingdom still has the largest number of ISO 9000 certified firms.

2. Fifty-seven percent of the respondents work for organizations with 250 or more employees, and only 20 percent worked for organizations with fewer than 50 employees. These percentages are interesting because they reveal that small firms are underrepresented. Indeed, U.S. government statistics reveal that 96 percent of all firms operating in the United States have fewer than 50 employees. Given that the majority of firms operating in the world are small businesses, one is surprised to learn that only 20 percent of ISO 9000 registered firms are small companies; one would have expected a larger percentage. (ISO 9000 registration is a greater burden for small businesses.)

   At any rate, if the percentages are truly representative, the survey indicates that ISO 9000 certification is still mostly achieved by medium to large companies. Finally, if one is to assume that a similar percentage breakdown would apply to other industrial countries, one could also conclude that the survey represents the point of view of about 5 percent of all firms.

3. Fifty-six percent of the respondents are quality managers. If one adds the two categories for "auditor" and "quality consultant" (6 percent each), we see that 68 percent of the respondents represent the quality or auditing profession. A mere 16 percent are line managers, business managers, or executive managers. One would like to know if these individuals share the same values regarding the ISO 9000 series as the other 68 percent. I personally doubt it.

4. Given the obvious, but no doubt unintentional, bias of the survey, one is not too surprised to learn that 88 percent of the respondents would like to see ISO 9000 and ISO 14000 changed. However, a consistent reply to any question does not necessarily mean that the question is meaningful or that respondents are in unanimous agreement. As Stanley L. Payne demonstrated long ago in his *The Art of Asking Questions*, "The more meaningless a question is, the more likely it is to produce consistent percentages

when repeated."[1] The survey question is itself odd because it states: "Should the ISO 9000 and ISO 14000 standards be left unchanged?" One wonders why both standards are mentioned in the same question. Why not a separate question for each standard? And yet, although many of us, including real customers, would like to see the ISO 9000 series of standards improved (changed) this does not mean that we would like to see it change to the current revision of ISO 9001-2000.[2]

Should All Customers Be Treated Like Kings?

Quality gurus, consultants, and even business owners often like to tell their audience that the customer is always right because the "customer is king." But is that really true of all customers? Perhaps, but one should also recognize that not all kingdoms are of equal importance or influence. Certainly, customers should be treated with respect, but, as this next story proves (provided by a friend of mine thanks to the wonders of the Internet), not all customers are created equal and therefore, cannot be treated equally. What follows is "actual" dialogue of a former XYZ Customer Support employee:[3]

"Ridge Hall computer assistant; may I help you?"

"Hi, my name is [Bill] and, well, I'm having trouble with XYZ."

"What sort of trouble?"

"Well, I was just typing along, and all of a sudden the words went away."

"Went away?"

"They disappeared."

"Hmm. So what does your screen look like now?"

"Nothing."

"Nothing?"

"It's blank; it won't accept anything when I type."

"Are you still in XYZ, or did you get out?"

"How do I tell?"

"Can you see the 'C' prompt on the screen?"

"What's a sea-prompt?"

---

[1] Stanley L. Payne, *The Art of Asking Questions*, Princeton University Press, Princeton, 1951, p. 17. One could say the same about meaningless or difficult-to-interpret paragraphs written in international standards.

[2] A review of Question 3 (User category) reveals who these mysterious customers are. The question allows for six categories: (1) industry (the true user of the standard), (2) certification body (enforcers of the standards), (3) standardization body (the organization that generates but does not use the standard), (4) accreditation body (not a user of the standard, their customers are the certification bodies), (5) public services (few of these organizations use the standards), and (6) other.

The classification is very revealing because when one recognizes that the "public services" and the "other" categories are likely to play a minor role as users of the standard, one observes that three out of the other four categories are not really users of the standards, certainly not in the sense of category 1, the industry. Why is that so? Why would the survey designer want to include certification, standardization, and accreditation bodies as part of this supposed "customer" survey?

[3] I have replaced the name of the software (a well-known word processing software) with the name XYZ.

"Never mind. Can you move the cursor around on the screen?"

"There isn't any cursor; I told you, it won't accept anything I type."

"Does your monitor have a power indicator?"

"What's a monitor?"

"It's the thing with the screen on it that looks like a TV. Does it have a little light that tells you when it's on?"

"I don't know."

"Well, then look on the back of the monitor and find where the power cord goes into it. Can you see that?"

". . . Yes, I think so."

"Great! Follow the cord to the plug, and tell me if it's plugged into the wall."

". . . Yes, it is."

"When you were behind the monitor, did you notice that there were two cables plugged into the back of it, not just one?"

"No."

"Well, there are. I need you to look back there again and find the other cable."

". . . Okay, here it is."

"Follow it for me, and tell me if it's plugged securely into the back of your computer."

"I can't reach."

"Uh-huh. Well, can you see if it is?"

"No."

"Even if you maybe put your knee on something and lean way over?"

"Oh, it's not because I don't have the right angle—it's because it's dark."

"Dark?"

"Yes. The office light is off, and the only light I have is coming in from the window."

"Well, turn on the office light then."

"I can't."

"No? Why not?"

"Because there's a power outage."

"A power . . . a power outage? Aha! Okay, we've got it licked now. Do you still have the boxes and manuals and packing stuff your computer came in?"

"Well, yes, I keep them in the closet."

"Good! Go get them, and unplug your system and pack it up just like it was when you got it. Then take it back to the store you bought it from."

"Really? Is it that bad?"

"Yes, I'm afraid it is."

"Well, all right then, I suppose. What do I tell them?"

"Tell them you're too stupid to own a computer."

The word "stupid" is too harsh and certainly not appropriate for good customer relations, but still I believe the individual at XYZ's customer support is doing customer Bill a favor.

Should the management of XYZ try to capture the market represented by customer Bill or should XYZ simply accept that in some cases, it is better not to have customers such as Bill? I suppose the answer depends on your customer–supplier philosophy. And yet, irrespective of your philosophy, you should always remember that customers may be king, but their kingdoms come in various sizes. If you believe that all customers are always right, you will probably devote much energy and thus resources and money to try to retain customer Bill. The other point of view, which is the one I would favor, is to recognize that in some cases, it is better to let the competition deal with customer Bill. Let them invest the resources.

CHAPTER 7

Fads, Incompetence, Ignorance, and Stupidity

The first and great object to be aimed at by the Manager of East Mill is PROFIT.

—*William Brown*[1]

Introduction

The original title for this chapter was to be "On Stupidity." However, after some reflection, I decided that perhaps not all acts that appear to be stupid are indeed stupid, so I had to broaden my scope and assume that some of the events that I perceived as examples of plain stupidity were perhaps due to incompetence or ignorance. Of course, to commit what may be thought of as a stupid act does not necessarily mean that one is stupid because, if that were so, everyone could be called stupid. As the cartoonist Scott Adams once wrote, "We are all idiots except some of us don't ever recognize it!" As the entries listed in the accompanying boxed material clearly demonstrate, acts of plain idiocy are frightfully prevalent in corporate America.

The last example in the list is a good example of how good intentions can, in the hands of enthusiastic simpletons (or Grafaloons as Kurt Vonnegut would call them), easily be pushed to the edge of moronic philosophy. As Burton Malkiel once observed (referring to the insanity found on Wall Street): "Stupidity well packaged can sound like wisdom."[2]

Nor should we assume that acts of absurdity are only limited to American corporations. In South Korea, many banks are spending considerable sums to train their staff on how to smile at customers. The logic behind the training is that in order to be more competitive in this global economy, South Korean bank

---

[1]Example of an early mission statement taken from a manuscript kept from 1818–1823 by William Brown, a Dundee flax spinner. Quoted in Sidney Pollard, *The Genesis of Modern Management*, Harvard University Press, Cambridge, MA, 1965, p. 253.

[2]Burton G. Malkiel, *A Random Walk Down Wall Street*, W. W. Norton & Company, New York, 1990, p. 73.

Corporate Insanity or Fun Corporate Communications?

1. As of tomorrow, employees will only be able to access the building using individual security cards. Pictures will be taken next Wednesday and employees will receive their cards in two weeks.
2. What I need is a list of specific unknown problems we will encounter.
3. How long is this Beta guy going to keep testing our stuff?
4. E-mail is not to be used to pass on information or data. It should be used only for company business.
5. This project is so important, we can't let things that are more important interfere with it.
6. Doing it right is no excuse for not meeting the schedule. No one will believe you solved this problem in one day! We've been working on it for months. Now, go act busy for a few weeks and I'll let you know when it's time to tell them.
7. My boss spent the entire weekend retyping a 25-page proposal that only needed corrections. She claims the disk I gave her was damaged and she couldn't edit it. The disk I gave her was write protected.
8. Quote from the boss: "Teamwork is a lot of people doing what 'I say.'"
9. My sister passed away and her funeral was scheduled for Monday. When I told my boss, he said she died so that I would have to miss work on the busiest day of the year. He then asked if we could change her burial to Friday. He said, "That would be better for me."
10. We know that communication is a problem, but the company is not going to discuss it with the employees.
11. We recently received a memo from senior management saying, "This is to inform you that a memo will be issued today regarding the subject mentioned above."
12. One day my boss asked me to submit a status report to him concerning a project I was working on. I asked him if tomorrow would be soon enough. He said, "If I wanted it tomorrow, I would have waited until tomorrow to ask for it!"
13. As director of communications, I was asked to prepare a memo reviewing our company's training programs and materials. In the body of the memo one of the sentences mentioned the "pedagogical approach" used by one of the training manuals. The day after I routed the memo to the executive committee, I was called into the HR Director's office, and was told that the executive VP wanted me out of the building by lunch. When I asked why, I was told that she wouldn't stand for "perverts" (pedophiles?) working in her company. Finally he showed me her copy of the memo, with her demand that I be fired, with the word "pedagogical" circled in red. The HR Manager was fairly reasonable, and once he looked the word up in his dictionary and made a copy of the definition to send to my boss, he told me not to worry. He would take care of it. Two days later a memo to the entire staff came out, directing us that no words which could not be found in the local Sunday newspaper

could be used in company memos. A month later, I resigned. In accordance with company policy, I created my resignation letter by pasting words together from the Sunday paper.

14. This gem is the closing paragraph of a nationally circulated memo from a large communications company: "XXX Technologies is endeavorily (sic) determined to promote constant attention on current procedures of transacting business focusing emphasis on innovative ways to better, if not supersede, the expectations of quality!" This memo represented the epitome of meaningless jargon produced by one or more individuals who obviously have attended one too many quality awareness seminars.

employees must learn to smile while working. (Apparently work and smiling are two mutually exclusive activities in Korean banks.) Although the link between smiling and international competition is tenuous at best, one can understand that if a Korean bank operates overseas the need to learn to smile at customers may indeed be valuable. However, one wonders why Korean tellers working in Korea must now smile at their compatriots who are likely to see the practice as odd and perhaps even impolite or disrespectful!

On Stupidity

Few authors have written on the subject of stupidity. Of the few that have, one must note Paul Tabori and, more recently, the cartoonist Scott Adams. Two Czech novelists, Jaroslav Hasek and Franz Kafka, have also written about stupidity from very different perspectives.[3] Hasek pokes fun at the absurdity of the turn-of-the-century Austro-Hungarian military bureaucracy in his immensely popular novel, which describes the adventures of the Good Soldier Svejk (pronounced Shvayk). Hasek tells us Svejk "had left the military service ... after having been finally certified by an army medical board as an imbecile." For the next several hundred pages Hasek, who obviously had no great love for turn-of-the-century Austrian military bureaucracy and bureaucracy in general, masterfully illustrates with great humor who really are the idiots in Svejk's society. In a more somber but equally effective style, Franz Kafka describes similar scenarios of bureaucratic inefficiency and absurdity. *The Castle* and *The Trial* are two of Kafka's best works.

One wonders how much more successful the good soldier Svejk would have been than Kafka's protagonist named K, when confronted with the examples of bureaucratic nonsense and ineptitude discussed next.

---

[3]Paul Tabori, *The Natural History of Stupidity*, Barnes & Noble, New York, 1993; Jaroslav Hasek, *The Good Soldier Svejk and His Fortunes in the World War*, Penguin Books, New York, 1981; Franz Kafka, *The Castle*, Schocken Books, New York, 1998; Scott Adams, *The Dilbert Principle*, Harper-Business, New York, 1996. Other books by Kafka would include *The Trial*.

Federal Nonsense

One of the ultimate examples of embarrassing procedural nonsense is form I-94W, a visa waiver form, which must be filled out by all tourists prior to entering the United States. Fortunately for American tourists, foreign countries do not yet require tourists to fill out something equivalent to I-94W, perhaps because they perceive the form to be too absurd. A Brazilian friend of mine once told me that he was sure that no one had ever answered "yes" to any of the questions listed on form I-94W. Indeed, I once recall watching a tour guide telling his bewildered audience to simply check the "No" box for all questions. What types of questions could provoke such a reaction?

Under the Heading "Public Reporting Burden," one can read that the estimated average time to fill out the form is 6 minutes. This estimate assumes that all tourists have an excellent knowledge of English. But one wonders how many tourists would really understand the long convoluted sentences and words such as "turpitude"! The form concludes with the following statement: "If you have comments regarding the accuracy of this estimate, or suggestions for making this form simpler, you can write to INS. . . ." I would suggest that the INS eliminate the form. Think of the tens of millions of minutes that could be saved by tourists visiting our country every year!

The Case of the Truck Rental Agency

The following incident occurred on two separate occasions during a period of only 4 weeks. In the summer of 1997 I decided to rent a truck from one of the leading truck rental agencies in the United States. Knowing that summer months are always very busy months for rental agencies, I decided to make my reservation 6 weeks prior to my date of departure. As I concluded the transaction, I was assured that a 26-foot truck would be waiting for me at a specific time at a specific location (a gas station near my house). The day *before* I was supposed to pick up the truck, I stopped by the gas station to make sure that they had my reservation. Sure enough, everything was in order, the attendant even had my name written down. I was surprised and impressed.

The next day (Friday) I arrived at the gas station at around 8:45 a.m. ready to pick up "my truck." I gave my confirmation number to the attendant and waited for my truck. A few minutes later the attendant came back and informed me that they could not find my reservation and, worse yet, they had no trucks available. After some 30 minutes of calm negotiation, I was told that I could have a truck "as soon as possible," but no one could tell me what that meant (maybe by noon or 1:00 p.m. or later).

By 10:00 a.m. I was getting annoyed and decided to call the rental agency's customer service department. I wanted to know why "my truck" was not available at the promised date and time. To my surprise, and contrary to what I was told by the reservation office, I was informed that the company could not promise delivery for a particular hour! Amazed at the reply I asked the representative if a Friday delivery meant any time between 00:01 and 23:59. The reply was even

Welcome to the United States

I-94W Visa Waiver Form (partial list of questions)

A. Do you have a communicable disease; physical or mental disorder; or are you a drug abuser or addict?

Yes No

B. Have you ever been arrested or convicted for an offense or crime involving moral turpitude or a violation related to a controlled substance; or been arrested or convicted for two or more offenses for which the aggregate sentence to confinement was five years or more; or been a control substance trafficker; or are you seeking entry to engage in criminal or immoral activities?

Yes No

C. Have you ever been or are you now involved in espionage or sabotage; or in terrorist activities; or genocide; or between 1933 and 1945 were you involved, in any way, in persecutions associated with Nazi Germany and its allies?

Yes No

D. Are you seeking to work in the U.S.; or have you ever been excluded and deported; or been previously removed from the United States; or procured or attempted to procure a visa or entry into the U.S. by fraud or misrepresentation?

Yes No

E. Have you ever detained, retained or withheld custody of a child from a U.S. citizen granted custody of the child?

Yes No

F. Have you ever been denied a U.S. visa or entry into the U.S. or had a visa canceled?

Yes No

If yes, when?_____where?_____

G. Have you ever asserted immunity from prosecution?

Yes No

more incredible: "Yes, sir, that's what it means!" Of course, such ludicrous terms could not be found written anywhere in the contract.

One month later a nearly identical incident occurred with the same rental agency and the same gas station; however, this time, "my truck" miraculously arrived within 45 minutes. What I suspect happened was that someone else got "my truck" and I probably got somebody else's truck. First come first served apparently takes precedence over reservations. Obviously, management does not seem to be aware of any problem—or does not wish to implement corrective actions or corrective action follow-up procedures. This is understandable because no one ever asked me to report/record my complaints. As far as the rental agency is concerned, the problem never occurred! All must be fine and customers are obviously really happy since they apparently rarely, if ever, write to complain.

Partial Quality and the French Public Transportation System

Reading one of the web pages of the French Association for Standardization (AFNOR), I once came across an interesting announcement which stated that a bus line, line 115, "Porte des Lilas-Chateau de Vincennes," received the NF ("norme francaise" or French standard) mark for Service.[4] The announcement explained with some pride that the RATP (the Parisian Public Transport Network) was the first such organization to commit itself to quality. (I suppose quality of service was not an important issue for the RATP prior to the issuance of the NF mark.) We are told that the principal characteristics monitored by the NF Service marking are:

· Information provided to passengers before and during their travel (I am curious as to what this information might be besides timetable schedules.)
· Welcoming (I assume that this means that the driver must now say "Hello" to all passengers.)
· Regularity and punctuality at (scheduled) stops (I don't know how the bus driver will be able to control Parisian traffic.)
· Comfort (With respect to what?)
· Cleanliness (Good to know.)
· Security (One wonders how this can be achieved without assigning a policeman on each bus or each bus stop.)

Although line 115 is the only one currently NF certified, the information bulletin informs the reader that other bus lines will eventually apply for the prestigious NF Service mark. Meanwhile, I suppose Parisians will have to transfer from the one NF-accredited bus line to a non-NF-accredited line. I wonder if they will really notice the difference.

---

[4]This is not too surprising from a country where we find ISO 9000 certification for ski lifts and ski resorts, certified gas stations, and even French standards for tennis courts (XP p 90-110, "Sols Sportifs—Terrains de Tennis—Conditions de Realisation et d'Entretien"). Source: AFNOR web page as of June 1998.

Can You Sell Less Quality?

With the notable exception of the maverick Dr. Juran who has often been critical of management fads, few authors have ever written about the absurdity, often bordering on stupidity, of the various management quality fads. In fact, if one wants to read about managerial stupidity in general including quality management, one must refer to probably one of the greatest gurus of all time: the cartoonist Scott Adams and his famous cast of characters, which includes the main character Dilbert and a cast of associates. The political cartoonist Garry Trudeau—creator of Doonesbury—has also contributed some thoughtful cartoons on absurd managerial fads. In one such cartoon, Trudeau diverts his attention to one of his favorite topics: higher education. The scene is commencement day at Walden University. The president of the university is addressing the graduating class of 1998:

So how would I describe the performance of the class of '98? In a word, appalling. In the main, you arrived poorly prepared, undisciplined and incurious. Once here you treated your teachers with incivility and your schoolwork as an inconvenience. Your only goals were credentialing and a good time. So are we as educators demoralized? Not at all. Here at Walden we pride ourselves on our ability to accommodate declining standards! In a changing world, this college strives to be nimble and adaptable. You, the consumer, told us that you wanted less for your college dollar, and we listened to you![5]

The references to "declining standards" and listening to the voice of the customers-students who demand less for their dollars is particularly amusing, sadly accurate, but never explored by quality gurus. One wonders how familiar Trudeau is with the many attempts by various universities to implement total quality programs. Yet, the scenario described in Trudeau's cartoon is far from being fictitious. A nearly identical scenario was described by the author in his *ISO 9000 and the Service Sector*.[6] The jump from stupidity to fads (or vice versa) is both small and effortless.

Are Benchmarks Always Conducted to Better Serve the Customer?

Benchmarking is the practice whereby a company is supposed to survey the "best practices" of industry leaders. Some quality gurus who firmly believe in the benchmarking process often speak of the need to benchmark to achieve so-called "world-class" status. One of the problems with this theory is that only a few companies can ever achieve world-class status because if most companies could achieve world-class status then "world-class status" would have to be down-

---

[5]Garry Trudeau cartoon in *The Press-Enterprise*, May 10, 1998.
[6]James Lamprecht, *ISO 9000 and the Service Sector*, Quality Press, Milwaukee, WI, 1994, p. 135.

graded to standard status. Of course, some gurus would likely retort that thanks to benchmarking some ethereal overall performance has been raised.

Regardless of whether or not this is true, quality professionals seem to ignore that companies can actually use the benchmarking process to reduce the overall quality of a product or service rather than improve quality. Airlines are a good example of an industry that has learned to benchmark down, that is, offer less while at the same time increasing prices. (Hospitals have also learned the practice of less service for more money.)

During the 1990s, I traveled extensively in the United States and abroad. One thing I noticed during the late 1990s is the reduction in service and the general increase in price within the American airlines industry. One of the industry leaders that has mastered the art of offering less for more seems to be American Airlines. The airline used to offer peanuts and, on 3-hour or longer flights, lunch or dinner including several rounds of drinks (soft drinks or, for a price, beer or other liquors). Slowly, the company began to shift from peanuts to pretzels, which taste more like cardboard but are, according to flight attendants, exactly what calorie-conscious travelers desire. Next came the reduction in drinks. Once, on a 4-hour flight from Dallas to Seattle, the author was served only one small glass of a soft drink and had to ask for another glass 2 hours later. On another flight, when I asked to have the can (because I was thirsty), I was told to wait because the flight attendant was not sure she had enough drinks for everyone!

Over the years I also began to notice that meals were systematically reduced in size. Lunch, when offered, consists of the same chicken sandwich or pasta salad with potato chips, a cookie, and a (free) glass of water or soft drink. Dinner is not much more. Quantity and quality have generally gone steadily down; meanwhile the average prices have gone steadily up and overbooking has become a standard practice. As if all of these cost reduction practices were not enough, some airlines now charge $5 for renting headphones to watch a movie!

Most airlines have learned that most passengers are actually willing to be abused (see comments on relativity of quality in Chapter 3). Nowadays all American airlines try to avoid offering lunches or dinner, and if they do, the meals all look alike and are equally bland. Fortunately, not all airlines charge the same prices and with some shopping around one can save as much as $1200 or slightly more on a flight from San Diego to Montreal (hard to believe that the airline business is a competitive business). What is remarkable about these practices is that more and more airlines have begun to offer less service for more money; they probably benchmarked what American Airlines was doing and they liked what they saw so much that these "best practices" are now part of the American airline industry. (Fortunately, to my knowledge, no foreign airlines have yet adopted these American best practices; perhaps they will soon.[7])

---

[7]Unfortunately, some Mexican airlines (Mexicana, for example) have already adopted the practice of overbooking.

On Quality Fads

Probably one of the best methods for following trends in quality or management issues relating to quality is to read *Quality Progress*, the official publication of the American Society for Quality (ASQ). As of early 1998, one of the fads that seemed to be on the ASQ's agenda was to recycle the six-sigma methodology, made popular in the early 1990s and apparently in need of being resuscitated. The May, June, and July 1998 issues of *Quality Progress* featured full-page advertisements for "The Vision of Six Sigma Breakthrough." The advertisement urged the reader not to miss this opportunity (apparently of a lifetime): "Don't miss the chance to be *certified* as Black Belt SM before year-end" the advertisement implores.[8] Just in case the reader might have missed the promotional page, the editor made sure that the June issue of *Quality Progress* had an article praising the virtues of six sigma: "Six Sigma and the Future of the Quality Profession."[9]

Of course, as everyone (including consultants) knows, no one program can save any company from the vicissitudes of the global economy. Even companies such as Motorola, which has transformed six sigma into a religion, have had to announce the layoff of 15,000 people because of the Asian economic crisis that began in October 1997.[10] Nonetheless, one must recognize that methodologies such as six sigma can at least help a company delay, and perhaps on occasion even avoid, economic disaster.

Besides cartoonists such as Adams and Trudeau, few authors and/or publishers have been brave enough to criticize the plethora of quality and associated managerial fads. The few I am familiar with would include Richard Farson's *The Management of the Absurd* (Simon and Schuster, 1996), Philip K. Howard's *The Death of Common Sense* (New York, Random House, 1994), Michael McGill's *American Business and the Quick Fix* (New York, McGraw-Hill, 1988), Peter Block's *Stewardship* (San Francisco, Berrett-Koehler Publishers, 1993), and Eileen Shapiro's *Fad Surfing in the Boardroom: Reclaiming the Courage to Manage in the Age of Instant Answers* (Addison-Wesley Publishing Company, 1996). Miss Shapiro includes a wonderfully witty dictionary for fad surfers. The dictionary alone is well worth the purchase of the book. For instance, "quality" is defined as "1: A recent rallying cry used by many, defined by few, and seldom the basis of thoughtful discussion; 2: (rarely) Superiority as defined by what the customer is willing to pay for."[11]

Although each of the authors just mentioned offers some thoughtful comment (and I do recommend their books), most of what they cover was already

---

[8]*Quality Progress*, June 1998, p. 103, emphasis added.

[9]Roger W. Hoerl, "Six Sigma and the Future of the Quality Profession," *Quality Progress*, June 1998, pp. 35–42. The January 2000 issue of *Quality Progress* published four letters (pp. 13–15) criticizing six sigma and the obvious promotion by the periodical. Could it be that the membership is finally awakening?

[10]"Motorola: Bad to Worse?" *USA Today*, July 7, 1998, p. B1.

[11]Eileen Shapiro, *Fad Surfing in the Boardroom: Reclaiming the Courage to Manage in the Age of Instant Answers*, Addison-Wesley Publishing Company, Reading, MA, 1996, p. 222.

addressed by Michael McGill and, to some extent, by Dr. Juran.[12] McGill critically reviews a broad range of managerial fads from the 1950s to the 1980s: organizational structures, management by objective (Drucker), specialized techniques (PERT/CPM) and psychological employee profiles (1950s) (reintroduced during the 1980s by some consulting firms), operation research, quantitative management, sensitivity training, matrix management, MBA training programs (1960s but still popular today), one-minute managers (1970s and 1980s), and so on.

McGill points out that although most of these fads were discovered 40 years earlier they were easily reintroduced to a new breed of managers. As McGill explains, all that is needed for a fad to be successful is to reduce everything to simple truth and "acronymic formulas": MBO, PERT, 9-9 management (one might add ISO 9000, TQM, SPC), and so on.[13] Once a fad becomes successful because it is simple and thus true, the books, videotapes, audiocassettes, and diskettes or CDs soon follow. The fad promising effortless implementation is now well established and has reached its maturity stage. The "quick-fix" package has achieved its objective and management is happy at least for a short while. A new fad is about to be born.[14]

In a tone similar to Richard Farson's (*Management of the Absurd*), McGill concludes that:

> *Fads and fixes emerge from the myths of management and, in turn, promote the same myths. The result is a managerial morass wherein simplistic solutions take form, flower briefly, then sink back to feed new forms. So long as management stays mired in its own mythic-ridden marshland, effective substantive solutions will evade managers, trapped as they are by their own imagined constraints and by objectives chosen by habits.*
>
> *If anything, there is an anti-science, anti-intellectual sentiment among managers today. In part, this is because the very processes of science—the deliberate (slow) search for specific (not universal) alternatives (not solutions) to be administered by experts (not managers) in complex (not simple) situations—flies in the face of what managers want.*[15]

As George Zipf noted several decades ago, "When confronted with a variety of pathways to an answer, people choose the one that requires the least amount of work." In other words, as Clifford Stoll observes, "People are lazy. Ease of use is more important than content."[16] The popularity of "how to" books clearly proves Zipf and Stoll right. More recently, the popularity of so-called "ISO 9000 software packages," which are used to miraculously "implement" an ISO 9000

---

[12]Michael McGill, *American Business and the Quick Fix*, McGraw-Hill Book Company, New York, 1988.

[13]Cartoonist Scott Adams has rediscovered the humor of this reductionism philosophy.

[14]McGill, *American Business*, p. 26.

[15]Ibid., pp. 5–6, 31–32.

[16]Zipf's comments quoted in Clifford Stoll, *Silicon Snake Oil*, Doubleday, New York, 1995, p. 184.

TABLE 7–1 How Would You Rate the Seminar and the Speaker?

	Seminar	Speaker
Number of responses	287	287
Minimum score	4	4
Maximum score	10	10
Mode (most frequent)	8	9
Average (mean)	8.03	8.4
Standard deviation	1.22	1.81

quality assurance system (and even have it certified), demonstrates that Zipf and Stoll are correct: the easiest path is the preferred path.[17]

Ignorance: The Leading Cause of Absurd Behavior

Some years ago I used to subcontract my services to a New York-based organization that offered, among its vast array of "quality" courses, ISO 9000 seminars. I have not seen an advertisement from the organization for some years. They may well be out of business by now, but not, however, before having earned millions of dollars from "their" ISO 9000 seminars. The remarkable thing about this and possibly other training organizations that were and still are in the lucrative "training business" was that the owners were not really interested in the quality of training but rather in how quickly they could offer a course and, of course, above all, how many people would attend. To attract as many people as possible, these organizations would rely on mass mailing of tens of thousands of flyers. With a mailing that large, even though the average rate of return is only 2 to 4 percent, it still means thousands of people took these ISO 9000 courses. At a cost of about $650 (1991–1992) for a 2-day course, one can imagine how much money was made just on the ISO 9000 courses.

Whenever I taught a seminar (and I offered about a dozen during 1991 and 1992 before I decided that it would be better to part company), I was required to distribute an evaluation questionnaire. This was and still is standard practice. Tables 7–1 and 7–2 summarize some 287 questionnaires collected between 1991 and 1992. Of the half a dozen or so questions listed on the questionnaire, I have selected the following two questions for discussion:

1. How would you rate the seminar?
2. How would you rate the speaker?

---

[17]In an excellent article written by Bert Gunter, the author expresses dismay at how some people seem to think that a quality assurance system can be implemented by simply purchasing a software package to produce "several pounds of paper." Bert Gunter, "Farewell Fusillade," *Quality Progress*, April 1998, pp. 111–119.

TABLE 7–2 Frequency Breakdown of Score for Seminar
and Speaker Questions

Speaker			Seminar		
Score	Frequency	%	Score	Frequency	%
4	3	1.05	4	3	1.05
5	4	1.39	5	9	3.14
6	8	2.79	6	18	6.27
7	43	14.98	7	42	14.63
8	73	25.44	8	116	40.42
9	112	39.02	9	71	24.74
10	44	15.33	10	28	9.76

All questions were to be rated on a 10-point modified Likert scale with 10 being the best score. [*Note*: The use of a 10-point scale is rather unusual; most Likert scales are based on a 5-point scale (1–5).]

I had assumed that my scores were fairly good. Only 5 percent of the population gave me a score of 4 to 6 and as much as 55 percent of the population gave me a 9 or 10. Lecturers who have offered courses know very well that it is impossible to satisfy all students-customers every time. A small percentage of students-customers, irrespective of one's efforts, will always be dissatisfied no matter what is done. (These are individuals who often think they know more than the instructor.) It is a well-known fact that the percentage of impossible to please "customers" varies anywhere from 3 to 5 percent. (*Note*: It is also well known that one should not try to satisfy this residual 3 to 5 percent. To do so often alienates the other 95+ percent of the population.)

In my case, a little over 4 percent of the customers rated me a 5 or 4. Not bad I thought. But I was wrong. One day I received a phone call from one of the administrators. The conversation soon drifted toward my evaluation scores. "What about my scores?" I asked.

"Well," the lady replied, "they are not as good as we would expect. We have very high standards of quality."

This last comment left me almost speechless because as most lecturers working for this agency knew, revenues rather than quality were the company's primary concern. "What do you mean my scores are not good enough?" I replied, already annoyed.

"Some of our instructors receive perfect tens."

"You mean to tell me that some of your instructors only get tens and not even one nine or an eight?" I retorted.

"That's right," the lady replied obviously not aware of the absurdity of her statement.

I tried to explain to the lady that perfect scores were impossible and against human nature but she would not listen or understand my argument. She was convinced or feigned to be convinced that there was nothing wrong with perfect scores (after all, that was a mark of excellence). I reminded her that since the questionnaire was administered by the instructors perhaps some instructors were removing scores they did not like and only submitting perfect scores. (A rather stupid thing to do but not necessarily that stupid when you are evaluated by someone who is more stupid than you are.) She would not listen. I realized then that it was time to resign. I stopped subcontracting with the organization within a few weeks.

Obviously, the lady had never taken a course in statistics and knew nothing about probability distribution. Repeated perfect scores (when dealing with random populations of evaluators) are a statistical impossibility or, at least, they represent events of very, very low probability. I never quite understood why I rated better than the seminar. The difference in score is actually statistically significant. People apparently liked the seminar but they liked me a little more!

This example of ignorance is not unique to training agencies. Ignorance bordering on stupidity can be found in most industries where individuals with a little knowledge reach beyond their abilities and wreak havoc among people who must satisfy their ludicrous requests. Indeed, the combination of power and authority and limited common sense or just plain ignorance often leads to an unfortunate lethal potion of irrational requests of the type so often celebrated by Scott Adams in his cartoons.

On Incompetence

Over the years I have come across some pretty interesting supplier evaluation questionnaires, but few have been as simple minded as the one I will now review. The questionnaire in question was uncovered while visiting a small shop of about 45 employees. The shop produces a sensor that is mounted on trucks and used to weigh loads. The unit can be used on logging trucks and sanitation trucks to weigh bins and charge customers according to weight rather than volume. The unit is so sensitive that it can also be mounted on beds, and is used by hospitals. It was in fact a hospital that had recently mailed to this shop a supplier quality survey, arbitrarily asking that the company be ready for ISO 9000 certification within 6 months.

The questionnaire consisted of more than 250 questions spread over 13 pages! The questions, inspired partly from the ISO 9001 standard, are typical and were probably copied from other surveys without even giving much thought as to their significance or relevance (a mistake often made by purchasing agents who thoughtlessly sent questionnaire clones). Naturally, since it is not practical to review all 250 questions, I have only selected the "best" 10.

Before reviewing these questions, a word about the scoring. The scoring is remarkably biased and poorly thought out in that the contributors have arbitrarily decided what is a correct or incorrect answer. For example, to the question "What is the annual number of suggestions per employee?," the supplier can score one point if he has an employee suggestion program (of one suggestion per year) or zero points if he has no such program. The obvious assumption is that you must have some program in place in order to score one point.

Some of my favorite questions follow:

1. Is your facility covered by ISO 9000 certification? (Two points for yes, one point for working on it, and zero otherwise. This question is not unreasonable except for the fact that the supplier already has an excellent rating and is rated as a preferred supplier. Also, the evaluator assumes that ISO 9000 certification is better than no certification. One would like to know on what quantitative information this belief is based.)
2. Who has received quality training? (Two points if the CEO and staff have received training, which explained the presence of a consultant teaching principles of TQM to the whole staff. Does anyone know what "quality training" is?)
3. How long are certificates and inspection records maintained? (One point is awarded if you keep records for at least 1 year! Why? Is it important to keep all inspection records for all products more than 1 year? Record retention will depend on the types of records and the nature of the business. For medical device manufacturers, FDA requirements for record retention are for the life of the product plus 2 years.)
4. Do control charts contain notes on process changes? (The requirement for including notes on process changes is reasonable; however, you are not given the option of answering "not applicable." You must answer "yes" or "no." The assumption is that having SPC charts is necessarily better in all situations.)
5. Is experimental design used? (yes = 1, no = 0) (Similar comments to question 4. One wonders if the person who drafted this questionnaire even knows what experimental designs are.)
6. Who is responsible for compliance? (Yes or No?) (Obviously an error; however, this indicates that no one probably edited the questionnaire.)
7. How is the maintenance department trained? (Two points for formal training plus on-the-job training, one point for on-the-job training, and no points for no training. I suppose previous experience does not count. Moreover, why should a customer even ask that question?)
8. Is just-in-time in use? (yes = 1, no = 0) (Note that the supplier is arbitrarily penalized for not using JIT. He can still deliver the parts on time but if he does not have JIT, he scores a zero!)
9. What is the average set-up time? (1 = ?, 0 = ?) (I suppose an "I don't know" answer is worth zero points. This question is impossible to answer since the choices are meaningless.)
10. Is process FMEA used for critical processes? (The acronym FMEA, which stands for failure mode and effect analysis, is not even defined! Why should the customer want to know? I have nothing against FMEA but can we first agree on what are "critical processes"? Who should define these "critical processes"? This question was probably lifted from one of the many automotive-type questionnaires. Is FMEA the only method? Is it the appropriate method? Is it the most cost-effective method? Shouldn't the supplier be allowed the freedom to select the appropriate method rather than have the customer dictate a solution?)

What do suppliers do when faced with such lengthy questionnaires? They usually hire a consultant to help them implement the system and move on hoping that the next customer survey will not add further requirements; unfortunately, the list of demands never seems to end.

The preceding questions are rather typical of the "Thou shalt do as I say, if you want my business" philosophy, which has been inherited from the military and the automotive industry (QS 9000). The perception is that, unless you are tough on your suppliers and demand that they implement all sorts of techniques (the more the better), they cannot possibly provide you with quality parts. In most cases, such questionnaires are rarely thoughtfully "designed." Often, they are simply plagiarized or are bastardized versions of someone else's question-naire. They are often the product of incompetent people who unfortunately have no idea of what they are requesting. Uncertain as to what they should request, they make ludicrous demands. Sadly, they are invariably unaware of the prepos-terous nature of their demands and the costs associated with their implementa-tion. Unfortunately, when combined with power or economic clout, incompetence is transformed into an evil destructive force.

The problem with incompetent people is that they tend to be very good at mindlessly following rules and regulations. As James Q. Wilson noted some years ago in his excellent book on bureaucracy, "[A] study of fifty-eight enforcement officials in forty-three municipal regulatory agencies found that those workers with the least training and talent were the ones most likely to take a legalistic view of their agency's rules and to apply them in a mechanistic fashion without regard to their overriding purpose."[18] These are the very people that are likely to end up on international committees demanding that more requirements and more guidelines be added to an already long list of standards.

On the Limitations of Mission Statements

Few activities have caused so much frustration and generated countless unpro-ductive meetings as the mission statement. A product of the 1980s, the mission statement has permeated almost every aspect of corporate and public life. Com-menting on the use of mission statements in public agencies, Wilson wrote: "If the agency's goal is so vague as to be meaningless (for example, 'advancing the interests of the United States') the administrator often will not know what to do and thus cannot be expected to tell the subordinate what to do, much less judge the work after the fact."[19] Often, the same is true in the private sector.

Many years ago the late French humorist Fernand Reynauld told the follow-ing story. A man is hired by a vendor to sell oranges at a stand in an outdoor public market. On his first day on the job our newly hired orange salesman notices that despite his various attempts to attract the attention of shoppers, no one bothers to even look at his oranges. He is just about ready to give up when he notices, hidden behind some rags, a slate and some chalk. "I've got it," he cried

---

[18]James Q. Wilson, *Bureaucracy: What Government Agencies Do and Why They Do It*, Basic Books, New York, 1989, p. 344. The same mechanical and thoughtless process of rigidly following rules and (worse yet) guidelines has been observed for decades among auditors for the Department of Defense, OSHA, and, more recently, some ISO 9000 auditors.

[19]Ibid., p. 156.

with joy, "I'll write a sign. But what to write?" After giving it some thought our salesman, who was no Shakespeare, writes the following words:

"Sold here beautiful oranges for cheap."

Proud of his newly acquired marketing skills, the man carefully places the sign atop his oranges. Moments later the owner happens to walk by the stand. He stops and carefully reads the sign aloud, pausing on every word: "Sold here beautiful oranges for cheap."

Owner:	What is the meaning of that sign?
Salesman:	I am trying to promote your oranges.
Owner:	Hum, what do you mean by "here," you are not selling oranges over there, are you? Why do you need the word "here"?
Salesman:	You are right, I did not think about that. The word "here" is kind of redundant, I'll erase it. (The salesman reaches for a rag and erases the word "here.")
Owner:	Are the oranges rotten?
Salesman:	No, of course they are very nice oranges.
Owner:	So why do you have the word "beautiful"? Customers can see that these are beautiful oranges.
Salesman:	You are right. I'll erase "beautiful."
Owner:	Do you intend on giving away your oranges?
Salesman:	No, of course not!
Owner:	So why do you need the word "sold"?
Salesman:	Yeah, that's true. (Proceeds to erase the word "sold.")
Owner:	Why say "for cheap"? These are very nice oranges, you don't want people to think that they are cheap oranges do you?
Salesman:	Of course not! (Erases the words "for cheap.")
Owner:	Are you selling bananas?
Salesman (without saying a word erases the last word on his sign and says to the owner):	I think I'll go sell fish.
Owner:	Why is that?
Salesman:	At least you can smell it.

Every time I read a mission statement I am reminded of the above story. All too often, mission statements tend to be nothing more than a series of meaningless words camouflaged as thoughtful sentences. The following mission statement is not the worst I have read, but one wonders why the owners bothered to have it printed on every paper cup:

To make great tasting, homemade food from the freshest, highest quality ingredients and to serve it quickly in a clean, friendly, and relaxed atmosphere.

I have not yet eaten in a dirty restaurant. Nor have I experienced a tense atmosphere in a restaurant—very noisy yes, but not to the point of causing a tension headache, for example. Quick service is nice but that may not be essential and may even slightly contradict the claim of a relaxed atmosphere. I have no way of knowing if the ingredients are the freshest and of the highest quality (nor I am sure the owner will ever be able to truly prove his claim). However, I have a sense of what fresh ingredients might be. While vacationing in Italy, my wife and I observed on more than one occasion local farmers delivering their produce to local restaurants. This produce certainly appeared to be very fresh (probably right from the field to the kitchen). Now, that is as fresh as you can get. I doubt that this restaurant has local farmers deliver their produce to the kitchen; maybe freshly frozen produce but not fresh produce. I am not sure why the claim to "homemade food" is made in the mission statement. The food is certainly not cooked at home; however, it does sound nice and so homey. Every employee is certainly very friendly.

Ever since the early 1980s quality gurus have often preached that in order to be successful an organization must have a mission statement. Mission statements are supposed to state a company's goals and objectives; this in turn is supposed to help employees focus on their role (mission) in helping the company achieve its goals and objectives. The problem with mission statements is not so much their purpose for they can be valuable, but rather, their blandness and triviality. Few mission statements (which are often confused with quality policy) ever say anything profound or of any value to a company or its employees. The cartoonist Scott Adams, who has often satirized the futile exercise companies go through to invent mission statements, has demonstrated how mission statements can result in meaningless sentences. While posing as an expert management consultant, Adams managed to convinced a group of executives who had no idea as to his real identity, to write the following nonsense mission statement: "The New Ventures Mission is to scout profitable growth opportunities in relationships, both internally and externally, in emerging, mission inclusive markets, and explore new paradigms and then filter and communicate and evangelize the findings."[20] It is truly remarkable that the executives approved such nonsense; no wonder Dilbert is such a popular cartoon character for millions of office workers throughout the world.

And yet, one must also acknowledge that with the proper level of commitment, some mission statements can make a difference. While walking through the charming historic district of Queretaro, Mexico, a city located in Central (Colonial) Mexico, I saw a sign painted on one of the walls of the Governor's Palace. I no longer recall exactly what the sign said, but it was a mission statement from the governor to the citizens of Queretaro. The statement was short and included a few promises about working together to improve the city. Based on what I witnessed in Queretaro, it is evident that the mission statement was not simple rhetoric. During my brief stay, I witnessed on more than one occasion, an army of men and women dressed in orange uniforms constantly sweeping the streets using brooms and even motorized vacuum machines. Leaves and other debris do not last long on the streets of Queretaro. A taxi driver assured me that during the past many years, Queretaro has consistently won the distinction of being

---

[20]"Dilbert Creator Pulls Off Consultant Hoax," *The Californian*, November 16, 1997, p. A7.

Mexico's cleanest city. No doubt, the governor's commitment to his mission state-ment is evidenced daily in the streets of Queretaro.

Side Effects of Exceeding Expectations

An often used sentence found in mission statements is this phrase: "We shall exceed customer expectations." Whenever I read such a sentence I cannot help but ask the company representatives how they know that their company actually does exceed their customers expectations. Do all customers have the same expec-tations? Are the customers' expectations documented anywhere? Because the answer to these questions is usually "no," one would like to know how an un-specified expectation can be exceeded? But one could go even further. I would propose that one of the most universal customer expectations is "Sell me your products or services for less." To exceed this universal customer expectation, a company would have to sell its products or services for less than expected or advertised! In fact, some customers would be perfectly happy (but perhaps not satisfied) if products could be had for next to nothing.

In some cases customer expectations cannot be exceeded because doing so would either violate a contract or add unnecessary cost to a product. For example, if a product must be delivered on a particular date and within a particular window of time, a supplier will not be able to exceed that expectation by delivering earlier than expected. To do so may actually result in some penalties being assessed by the customer. In the case of a supplier who is asked to deliver a product with a certain purity, say, 97 percent pure, exceeding the purity level would certainly exceed his customer's expectations but at what cost or benefit to the supplier?

Finally one must consider the case where exceeding a customer expectation may cause problems for another customer. This problem is not uncommon and can be attributed in part to the belief that the customer is a monolithic entity. Indeed, one often reads about the "voice of the customer" as if customers speak with one voice. Anyone familiar with the daily activities of any company would immediately point out that "the voice of the customer" is a myth; customers rarely universally agree on what they want (except for the cheapest price possible). The following anecdotes illustrate the danger of exceeding a customer expectation.

One of the basic raw materials used by the printing industry is ink. This par-ticular printer was receiving its ink from a very reliable supplier. One day, the process manager reported to the quality manager that they were having all sorts of problems with the ink (the ink would not dry as fast as it used to, creating many smears and thus rejects). What had gone wrong? The quality manager immediately contacted the supplier, explained the problem, and wanted to know if anything had been changed in the formula. The supplier assured the manager that she was receiving the same ink that had been delivered for the past several years.

After several days of back and forth phone calls, the ink supplier finally admit-ted that an important customer (a customer more important than the printer) had requested a small change in the formula that would improve the "quality" of the ink. The supplier was sure that the change could not possibly affect the printer; besides the product had been improved and exceeded the printer's specifications. Unfortunately, the improvement did create havoc for the printer. Improving or exceeding a customer specification may result in reduced performance for another customer.

In some cases, a supplier's refusal to admit that a specification was "improved" could lead to major difficulties lasting for several months. Such was the case for the quality manager of a computer manufacturer who suddenly found that his rejection rate for diodes jumped to an incredible 50 percent. It took nearly 4 months for the Japanese supplier to finally admit that they had made some changes to the diode to satisfy one customer but had failed to inform other customers. Meanwhile the Japanese supplier kept insisting that the problem must have been with the customer's in-process inspection test.

This last example concerns one of the leading manufacturers of ropes used mostly for maritime applications. The rope manufacturer purchases its synthetic fibers from one of the top two or three chemical suppliers in the country (all are ISO 9001 or 9002 registered). The supplier's main customer is the automotive industry, which purchases the bulk of its synthetic fibers for use principally in tires. When the automotive customers request specification changes in certain fibers, the supplier modifies its formula to satisfy its important customers. The rope manufacturer, which is a very small customer when compared to the automotive industry, learns of the changes only after the chemical supplier mails new advertising brochures listing the new specifications. These formula changes impact the rope manufacturer who must now change all machine settings to reflect the new specification. Letters of complaint to the chemical manufacturer have never led to anything.

When Too Much Quality Leads to Ludicrous Scenarios

In an article entitled, "Excuse Me . . . You Call This Service?," Del Jones noted that as employment rises so do customer complaints. Quoting the Bureau of Labor Statistics and the American Society for Quality, Jones writes that as unemployment went down from 6.5 to 5.0 percent (from 1994 to 1997), the American customer satisfaction index dropped (during the same period) from 74.2 to 70.7.[21] Coincidence or statistically significant correlation? We are not told, but the similarity in the negative trend is interesting. The following anecdote would certainly indicate that the quality of service is not what it used to be.

In July 1997, while trying to make some phone calls, almost every time I dialed a number, I got a busy tone or a recorded message. After dialing the number a second time I would usually get through. After a few days of this annoying problem, I decided to call the telephone company's customer service department. The lady who answered the phone said something most unusual and although I do not recall her exact words it was something like: "How may our service be of excellence to you, sir?" Not only was the expression odd (and likely to have been suggested by a high-priced quality consultant), but in view of all the problems I

---

[21]Del Jones, "Excuse Me . . . You Call This Service?" *USA Today*, July 22, 1997, pp. B1, B2.

had experienced a few days prior, I could not help but laugh and had to apologize to the lady. I explained my "new" problem and ended by asking when it would be fixed. The lady did not know of any problem and was sorry she could not help me. I then asked if she could transfer me to another phone number, perhaps maintenance, but she again offered her apologies and said she could not help. I thanked her and hung up the phone. So much for excellence in service.

About a week later, I received a call from the telephone company (GTE) customer service department. The lady explained that GTE was happy to have me as a new customer (GTE is the only service company in the region). She wanted to know how I would rate GTE's service: excellent, good, or poor. I replied: "I am sorry but I have to rate you as poor because I have had great difficulties with the installation of my new phone lines and the line has not been working properly over the past couple of weeks."

"I am sorry to hear that, sir" (the standard answer). She then said something very strange: "Thank you for helping GTE provide you with a better service. Is there anything else I can help you with?" I was so surprised that I thanked her and hung up the phone.

I find it incredible that the person could not even suggest another phone number or department. I could not help but wonder what is the purpose of customer service? I was not even asked when the problem occurred (day or time). The answer was simply "Sorry, I don't know." Would a procedure on how to assist customers be helpful in such instances? Perhaps, but I don't think the problem relates to a lack of procedures. Something more fundamental is needed: training.

The above stories demonstrate that procedures without training are worthless. I have no doubt that each of the persons who took my call were following some procedure(s). Some, for example, the lady who answered by asking me how she could provide me with excellent service, had probably been trained by consultants who thought that the slogan was clever and would help define some mission statement. No wonder that Dilbert is so popular. It would appear that someone forgot to define to the telephone operator in charge of customer services the meaning of excellence. So, without a clear understanding of what "excellence" means, the operator is simply left to parrot a statement obviously devoid of any meaning. Unfortunately, too many companies seem to believe that once they have written a mission statement and "trained" some employees on how to parrot a few statements, all is well.[22]

Perhaps the content of the training needs to be revisited. Indeed, the sociologist Robin Leidner who has studied the written scripts that are handed out to employees in service enterprises reports that "What these scripts aim to do is establish the 'friendliness' of the employee more than address the substance of the client's concerns. In a turnstile world of work, the masks of cooperativeness are among the only possessions workers will carry with them from task to task, firm to firm—these windows of social skill whose 'hypertext' is a winning

---

[22] I should explain that a few months after this story was written I received yet another call from customer service. This time service was impeccable.

smile. If this human-skills training is only an act, though, it is a matter of sheer survival."[23]

The Routine of Quality

The phrase "routine of quality" refers to the meaningless, desensitized, robot-like, standardized behavior that is experienced daily by millions of customers when they interact with a "customer representative." (Several examples have already been noted in previous chapters.) The nearest equivalent to this proce-duralized and mechanized behavior can be found in the equally meaningless "Have a nice day" statement that is invariably uttered at the end of most trans-actions between a customer and a retailer.

Examples of routine of quality would include recorded messages that inform the caller that the conversation may be recorded "for quality control." More and more businesses in the United States are using these preprogrammed recorded messages, which, unfortunately, do not guarantee that the customer/consumer will receive *intelligent* quality of service. What do I mean by intelligent service? Intel-ligent quality service is provided when an employee can listen to a customer com-plaint or suggestion and take action to adapt his answer to match the complaint rather than offer a standard answer written in some official company policy. Such service is offered by some companies but, invariably, the operator or individual taking the call must first verify with a supervisor before offering the "approved" answer.

The frequent inability of customer service employees to offer valuable assis-tance to customers may be in part attributable to the fact that these positions offer very little pay (and perhaps also inadequate training). A review of the "want ads" in the *Union Tribune* (a San Diego paper) revealed that most "customer service" jobs paid around $8.00 to $9.50/hour—not much financial incentive con-sidering the nature of the job.[24]

Conclusion

What is very strange and even surrealistic about the cases described above is that in many instances, the employees received so-called "quality training" but did not really know how to apply it or, worse yet, were not given the authority to use their training. (An attempt to correct this problem will be introduced in the new ISO 9001-2000 standard.) This was most clearly evident in the case of the GTE phone company. Everyone knows how to recite absurd statements but no one is given the authority to satisfy customers. In some cases, employees (or perhaps management) do not even know what satisfying the customer or even

---

[23]Leidner quoted in Richard Sennett, *The Corrosion of Character*, W. W. Norton & Company, New York, 1998, p. 112.

[24]One advertisement, offering $9.00/hour plus benefits, had "100 immediate openings." Another advertisement, offering a generous $11.75/hour, was looking for 25 applicants!

honoring a contract really means. And yet, every year hundreds of books are published about quality, management, and a myriad of related topics. It is as if there were three separate realities: the one written in books (theoretical), the one described in training sessions (when such are given; this is the attempt at teaching the praxis of quality), and last, but most important, the reality of everyday business life so wonderfully and humorously depicted in Scott Adams' comics. It is a wonder our society is not producing more schizophrenics.

The remaining chapters retrace the heritage of the quality control philosophy and review how the prescriptive practices of control exercised for the sake of quality improvement emerged as early as the 16th century when government agencies began to monitor manufacturing processes. An analysis of how these practices were eventually codified in international standards is also provided.

Part III

Colbertism and the Dawn of Power in Customer–Supplier Relations

CHAPTER 8

Colbertism: The Dawn of Regulatory Practices

The desire to protect consumers from the perceived or real evil practices of unscrupulous manufacturers has a long history. Examples of early inspection and audit practices designed to approve or certify products, have been traced back to the Pharaohs.[1] This long-established tradition of inspecting products for their quality and monitoring artisans and, later, manufacturers, is also observed in the middle ages. Indeed, during the last six centuries, the church, feudal lords, burgs, kings, and, more recently, governments have all played an important role in regulating or otherwise attempting to control the means of production. The late Belgian historian Henri Pirenne noted that, during the medieval ages, consumers were protected from fraud and adulteration by regulating industrial practices and supervising sales.[2] Pirenne proceeds to explain that

> *The discipline imposed on the artisan naturally aimed at ensuring the irreproachable quality of manufactured products. In this sense, it was exercised to the advantage of the consumer. The rigid regulation of the towns made scamped workmanship as impossible, or, at least, as difficult and as dangerous, in industry as was adulteration in food. The severity of the punishments inflicted for fraud or even for mere carelessness is astonishing. The artisan was not only subject to the constant control of municipal overseers, who had the right to enter his workshop by day or night, but also to that of the public, under whose eyes he was ordered to work at his window.[3]*

---

[1] J. M. Juran, "Quality Control under a Monopoly," *Industrial Quality Control*, March, 1965, p. 463. For an excellent historical overview of the use of the inspection stamp, particularly in the manufacture of firearms, see L. E. Snodgrass, "The Armorer's Mark of Quality," *Industrial Quality Control*, May 1966, pp. 586–591.

[2] Henri Pirenne, *Economic and Social History of Medieval Europe*, Harcourt, Brace & World, New York, 1937, p. 181.

[3] Ibid., p. 184. See also L. E. Snodgrass, "An Early Quality Control Standard," *Industrial Quality Control*, October 1962, pp. 34–35.

Corroborating Pirenne, Charles Woolsey Cole explains that from the 13th century on, the royal government began to take over functions that had formerly belonged to the church, the town, and other feudal authority. Recognizing the newly acquired power of the royal government, the voluntary association of guilds continued to seek the authority of a feudal noble, a bishop, or a king "for approval and ratification of their status."[4] By the beginning of the 13th century, the Republic of Venice's Arsenal, "that mighty complex of dockyards, foundries, magazines and workshops for carpenters, sailmakers, ropemakers and black-smiths,"[5] was already under government control and supervision. The monitoring of the Venetian shipyards, which had developed a kind of assembly-line manufacture, capable of producing up to 10 galleys per day, is noted by John Julius Norwich who explains that "most of the official controls on industrial standards and conditions of work were imposed through the guilds, whose carefully-framed statutes were all subject to the approval of the government supervisors."[6]

Historians agree that by the 13th century, royal enactments to control industry, similar to those found at the Arsenal, began to appear in Europe. By the end of the 15th century, a series of regulatory legislation dealing with the "preservation of quality" were issued by the King of France (1479, 1512, 1531, 1543, and continuously until 1584). In 1571 a royal ordinance setting standards of quality for dimensions of woolen fabrics and the "number of threads in the warp" was issued and, by 1586, royal inspectors were inspecting fabrics for the proper use of the royal seal.[7] By the 16th century, such legislation was common. It is worth noting that, although these regulations might have been issued to remove economic abuses, they also allowed the king to raise money to fund a nascent bureaucracy.

The effectiveness of such control and supervisory mechanisms, as exercised by city-states, such as the Republic of Venice, and the monarchies of the 14th and 15th centuries, is difficult to assess. Indeed, although the number of royal edicts increased exponentially from the 12th century on, and even though historians have documented their importance, it is nonetheless difficult to determine how rigorously these edicts were enforced throughout the kingdom. Based on the research of historians such as R. C. Cobb and George Rudé, one would have to conclude that as kingdoms were struggling to establish themselves as nation-states, the enforcement of royal edicts must have been a tedious and

---

[4]Charles Woolsey Cole, *Colbert and a Century of French Mercantilism*, Vol. II, Columbia University Press, New York, 1939, p. 365.

[5]John Julius Norwich, *A History of Venice*, Alfred P. Knopf, New York, 1982, p. 83.

[6]Ibid., pp. 84–85, 273. See also Edward R. Weidlein and Vera Reck, "A Million Years of Standards," in Dickson Reck (Ed.), *National Standards in a Modern Economy*, Harper & Brothers, New York, 1954, p. 13, where the authors mention that the standardization of weapons was already undertaken by the Arsenal. For a brief eyewitness description of the Venetian shipyards as noted by the Spaniard Pero Tarfur who toured them in 1436, see L. Sprague de Cam, *The Ancient Engineers*, Ballantine Books, New York, 1963, p. 385.

[7]Cole, *Colbert*, p. 366.

difficult process.[8] Indeed, it was not until the 17th century that government supervision on manufacturers was to be attempted on a national scale previously unheard of.

Colbertism: The Dawn of Modern Government Regulation

Jean-Baptiste Colbert (1619–1683), comptroller of finance in 1665 and secretary of state for Louis XIV in 1668, is responsible for introducing, over the span of a few years (1666–1669), a series of regulations aimed at controlling French manufacturers and designed to improve the quality of French goods. In reviewing "le système Colbert" as it was then known, I will demonstrate how Colbert's attempts at regulating French industry, noble as they were, had mixed results. The relevance for today's regulators is that, although the age of Colbertism was conceived more than 300 years ago, its lessons are timeless.

Overview of the French Economic System during the 17th Century

By the beginning of the 17th century, the French economy was essentially dominated by the activities of the state.[9] The emergent mercantilism is characterized by a series of regulations and tariff policies (1664) aimed at simultaneously protecting French industries from foreign competition and promoting French goods overseas.[10] French manufacturing consisted of essentially three types of manufacturers: a few "*manufactures du Roi*" or the king's manufacturers such as Gobelins (which still exists today) and Arsenaux. These manufacturers were state enterprises that deliver their products for a fixed, predetermined price.[11]

Next, one finds the "*royales manufactures*" which were founded by the king. The king favored these manufactures with various privileges and special favors such as the granting of monopolies. These manufacturers could use the king's

---

[8]For an interesting analysis of the evolution of the written document, see Ivan Illich and Barry Sanders, *The Alphabetization of the Popular Mind*, North Point Press, San Francisco, 1968. For an account of regional unrest and resistance to unification, see R. C. Cobb, *The Police and the People: French Popular Protest (1789–1820)*, Clarendon Press, Oxford, 1970, and George Rudé, *Ideology and Popular Protest*, Pantheon Books, New York, 1980.

[9]The following summary is based on Jacques Ellul's excellent *Histoire de Institutions de l'Epoque Franque à la Revolution*, Presses Universitaires de France, Paris, 1967, especially his Quatrieme Partie (La Monarchie Absolue) (XVIIe et XVIIIe Siecles), pp. 397–438.

[10]Coles (*Colbert*, p. 558) defines French mercantilism as "that group of theories, policies, and practices arising from the traditions of the country and the conditions of the time, and upheld and applied by Jean-Baptiste Colbert during his years in office, 1661–1683, in his effort to secure for the nation, and for the king who symbolized it, power, wealth, and prosperity."

[11]Braudel distinguishes between "manufactories" and "factories." Manufactory is reserved for the concentration of labor of an artisanal type using manual labor "and the word factory to enterprises using machinery such as was already to be found in mines, in metallurgy and in shipyards." By the 18th century, the two words were used interchangeably. (Fernand Braudel, *The Wheels of Commerce: Civilization and Capitalism 15th–18th Century*, Vol. 2, Harper & Row, New York, 1986, p. 329.)

mark on their products; however, they were privately managed and were in business to serve the general public.

The third and last type of manufacturer were "*manufactures privilegiés*" (mainly, printers, sculptors, painters, and inventors of industrial processes), for which their status of "privilege" simply indicated that they were exempt from corporate regulations.

In most instances, especially in the case of the textile and related industries, the manufactories were the focal point "of a network of cottage industry, the place where the production process was eventually finished."[12] In an attempt to better control their assembly processes, it appears that some manufactories such as Aubusson, for example—which employed 680 workers in 1789—relied on the use of work instructions and documented procedures (essentially diagrams and pictures since the workforce was illiterate).[13]

These merchants, who were invariably located in major cities, would "put out" work to hundreds or even thousands of subcontractors, and provide the artisan(s) with raw material and a part wage, the balance being paid on delivery of the product. Contrary to what has occasionally been assumed, these factories and manufactories often employed thousands of workers. Braudel, for example, cites Van Robais, in Abbevile, which employed 3000 workers; a stocking manufactory in Orleans, which in 1789 employed 800 persons; and a woolen factory in Linz, Austria, which employed an incredible 26,000 workers in 1775.[14]

Much as the U.S. defense industry, which burgeoned during the 1940s and blossomed in the 1950s and 1960s, was to rediscover, being a privileged manufacturer had its advantages and drawbacks. As Ellul observes

> For many of these manufacturers, the State practically becomes the sole client: it is the sole important purchaser of metallurgy. The State therefore circulates the bulk of the money in a closed circuit, maintained by the State: taxes, purchases by the State from manufacturers, payment for raw materials and salaries, from which taxes are raised.[15]

It is within this background that Colbert planned and implemented his system of industrial regulations.

---

[12]Ibid., p. 330.

[13]For examples see, Dominique Chevalier, Pierre Chevalier, and Pascal-Francois Bertrand, *Les Tapisseries D'Aubusson et de Felletin (1457–1791)*, Solange Thierry Editeur, La Bibliotheque des Arts, Paris, 1988.

[14]Braudel, *The Wheels of Commerce*, p. 332. Peter F. Drucker in his *The New Realities* (Harper & Row, New York, 1989, p. 221) seems to believe that 300 employees was rather unusual, even in the early part of the 19th century. The put-out system is still very much in vogue today. The software giant Microsoft, for example, subcontracts much of its programming work to hundreds of programmers. In many cases, the subcontractor is offered a work area and a computer to program.

[15]Ellul (*Histoire de Institutions*, p. 420): "L'Etat devient pratiquement le seul client de beaucoup de ces manufactures: il est le seul acheteur important de la métallurgie, Il fait donc circuler la masse monétaire en circuit fermé, maintenu par l'Etat: impots, achats par l'Etat aux manufactures, paiement des matieres premieres et des salaires, sur lesquels sont prélevés les impots." The significance of Ellul's observation will be explored in a later chapter when I describe a similar scenario that was to unfold nearly 300 years later with the emergence of the U.S. military as primary customer.

The Colbert System

The poor quality of French goods and the continued decline of French industry and productivity, which began at the end of the 16th century, was of great concern to Colbert. Assessing the situation, Colbert concluded that "French industry could be restored and built up, only if it were properly regulated."[16]

Colbert was convinced that it was chiefly by quality that manufactured goods could gain and retain market, both at home and abroad. Quality, he was sure, could be obtained only through the promulgation and enforcement by the central government of carefully drawn regulations. Such regulations would put the industries of the nation into a state of "good order," which he ardently desired.[17]

To achieve his objectives, Colbert's administration proceeded to publish between 1666 and 1669 numerous detailed regulations aimed at controlling most aspects of manufacturing and working conditions. Aware that neither he nor his staff had the required knowledge to draft industrial regulations, Colbert had the wisdom to consult with the leading industrials and merchants (i.e., the bourgeoisie) of his time. The set of regulations produced by the Colbert administration covered every aspect of the socioeconomic life of the time: hospitals; iron production (since it was already recognized that the "rupture of iron offered a danger to the public");[18] naval ordinances for ship building, because ships had to be certified to their capacity; working conditions; raw material specification for cannon balls and gun powder; manufacturing procedures for the dyeing of textiles; test procedures to be conducted on product; and in some cases even crude statistical experiments.[19]

Colbert's administration was well aware of the principles of continuous improvements as Article 56 of the Beauvais regulation of February 4, 1667, clearly demonstrates:

56. Every two months there is to be held, at the Episcopal palace, a general assembly for this industry. The maire, echevins, *wardens, former wardens, chief merchants, and manufacturers, and most skilled merchants and manufacturers are to attend. The purpose of the meeting is to improve, perfect, and*

---

[16]Ellul, op. cit., p. 421.

[17]Cole, *Colbert*, p. 363 (citing Colbert's Lettres and Correspondance Administratives).

[18]Cole, *Colbert*, p. 365.

[19]Maurice Daumas, for example, quoting Savary, reproduces one example of such manufacturing procedure: 'The good quality of this woad results from the fact that it is made with only six pounds of fully prepared indigo for each ball of woad, when the vat is low, that is, when the woad is beginning to give off its blue flower. . . . Next it must be boiled with alum, tartar or bitartar of potassium, and after that maddered with ordinary madder . . . and finally completed in black with nutgall . . . cuperas and sumac toned down, passing it over the yellowweed to give ot the perfection of black (etc.). In these instructions, Daumas concludes, "we find the essence of the 'dyers' method in the second half of the seventeenth century, as they were to remain for more than a century and a half." Maurice Daumas (Ed.), *A History of Technology and Invention: Progress Through the Ages*, Vol. II, *The First Stages of Mechanization*, Crown Publishers, New York, 1969, pp. 196–197. Coles (*Colbert*, p. 376) briefly describes how in 1682 a government official of Amiens by the name of Breteuil conducted a simple experiment to demonstrate that calendered and uncalendered *baracans* showed no superiority when dried in the shade or in the sun.

bring good order to the industry, to prevent abuses, and to send reports to M. Colbert.[20]

A unique feature of Colbert's set of regulations is that, in many cases, they were adopted to local needs. (As we shall later see, such foresight is not considered in today's international standards.) Unfortunately, Colbert's bureaucrats did not always take into consideration regional or local needs.

Having produced an impressive set of regulations, Colbert soon discovered that most of the regulations were not enforced. In the case of regulations on woolens, Colbert was advised by experts that the best thing to do "was to send skilled persons to inspect the various manufactures of the kingdom, and to regulate the length and width of the textiles made, the quality of the wool, and the methods of manufacture, and the relations between masters, journeymen, and apprentices."[21]

Colbert's system of inspection is worth reviewing because it demonstrates that the fundamental difficulties inherent in administering any national inspection/auditing program, were already well known before 1670.

Colbert's Rules for Inspectors of August 13, 1669

Of the 65 articles making up the rules for inspectors, and reproduced in Cole's book, I have only included articles 19, 30, 35, 46, 56, and 59. These articles, and many others like them, are interesting because, in principle, they approximate and indeed anticipate auditing procedures developed by the U.S. military during the early 1950s, which were repackaged/recycled in 1987 by the International Organization for Standardization, and currently practiced by ISO 9000 third-party registrars.

19. The inspectors are to see that the wardens make a note of the defects found and draw up suggestions for the improvement of the industry, and keep a register of all defective cloth discovered, so that the inspectors can see that workers are at fault.

30. Each inspector is to have a trustworthy agent who understands manufactures in every place where manufactures are established. These agents are to discover abuses and suggest ways of improving manufacturing.

35. The inspectors shall see that all weights and measures conform to the old ordinance. Any which do not are to be confiscated.

---

[20]Coles, *Colbert*, p. 381. Some of these regulations anticipated the zero-defect movement of the early 1960s by three centuries. Indeed, Boissonnade writes that, in the case of wool regulations, the primary objective was to force manufacturers to produce using the best technical rules, under the control of the state, solid wool with no defects. P. Boissonnade, *Colbert: Le Triomphe de L'Etatisme, La Fondation de la Suprematie Industrielle de la France, La Dictature du Travail (1661–1683)*, Marcel Riviere, Paris, 1932, p. 211. "La reglementation qui a pour objet d'obliger les fabricants, sous le controle de l'Etat, a fabriquer des lainages solides, exempts de defectuosites, conformes aux meilleures regles de la technique, indique les diverses prescriptions qui doivent etre obligatoirement observees."

[21]Coles, *Colbert*, p. 373.

46. . . . the inspectors are also to see that the dyers are properly equipped, and are to have a mark to put on the cloth they dye.

56. The inspectors should notice what places are suited to what dyes, because of the local water, herbs, roots, and leaves, "so as to oblige those of the surrounding country to send their stuffs there to be dyed," it being noteworthy "that good dyeing increases the quality, the beauty, and the price of their fabrics, although it costs little more than mediocre or bad dyeing."

59. The inspectors shall make a list of places where commerce and manufacturing are established, of what sorts, kinds, processes, and qualities. They shall secure samples and data as to the dimensions, prices, yearly production, and number of looms.[22]

Finally, inspectors were to make sure that redyed and refueled goods were re-marked and to ensure that working conditions were suitable.[23]

As is the case with today's third-party auditing, product inspection by royal inspectors was not free (2 sols 6 deniers per piece). This was know as "le droit de marque à la fabrique" or "the right to mark fabrication," and was very similar in intent to the many marks of quality issued throughout the centuries by various governments or government consortia, including the CE mark required by the European Community for certain regulated products. In addition, the inspector had to have at least 6 years of experience not only in the industry but also *in the manufacturing* of the product.

By 1700, it was recognized that since the inspectors had too broad a responsibility, including the establishment of fines for product deemed to be of poor quality, all candidates to the post of inspector had to now prove that they had special aptitudes to perform their duty.[24]

Problems with Colbert's System of Regulation

Although Colbert's regulations were less specific, less meticulous, and less severe than the older guild regulations, they were generally not welcomed by both employers and employees. Employees resisted the regulations because their job security soon became threatened. Indeed, relying on contemporary arguments, it did not take long for employers to argue that Colbert's costly regulations would cost jobs. Resistance to Colbert's bureaucrats and bureaucracy began in Paris (1666) and spread to Lyon (1667) and Nimes

---

[22]Coles, *Colbert*, pp. 421–424.

[23]Readers familiar with the ISO 9000 series of standards, to be reviewed in Chapter 12, will no doubt notice the similarity of these last requirements with paragraph 4.13, Control of Nonconforming Product, and paragraph 4.9b, Process Control.

[24]Germain Martin, *La Grande Industrie sous le Regne de Louis XIV, 1899*. Reprinted by Burt Franklin, New York, 1971, pp. 126, 327, 330. This requirement approximates current debates regarding the need to establish requirements for ISO 9000 auditors. It should be pointed out that Colbert's inspectors not only inspected facilities but also products.

(1682).[25] The following quotation from a soap manufacturer in Lyon, voiced in 1770, but equally valid today, epitomizes the general negative attitude toward inspectors expressed by manufacturers for the last 300 years:

> *A local inspector is charged with the inspection of all operations for each fab-rication. After the final operation, he inspects each sheet and marks them or rejects them. The only effect of these inspections can be reduced to: bureau-cratic cost, degradation of the merchandise, shipping delays, and the oppres-sion of the manufacturer who does not dare complain about the inspectors' injustices for fear that he might be subject to even more rigorous judgments.*[26]

The cost associated with Colbert's regulations had some unfortunate repercus-sions on other members of the socioeconomic system, namely, the thousands of rural subcontractors who eked out a living by subcontracting (i.e., "putting-out") their labor to the city merchants, many of whom belonged to the network of priv-ileged industries supported by Colbert. Colbert's policies, which were partly designed to encourage the concentration of industries within a few cities, were certainly not favorable to these small artisans.[27] To achieve as much geographic concentration of the economic system as possible, Colbert's administration wished to limit and even eliminate the "put-out" system for which it obviously did not have sufficient resources (i.e., inspectors) to control. Yet, although Colbert was only minimally successful in achieving his goal, the dispersion of industry still persisted by the end of the 17th century. However, as Dockés observes, the dis-persal of the artisan was only apparent because the cottage industries "were caught up in an invisible financial spider's web, its threads controlled by a few merchants."[28]

There were other fundamental objections to the principles of state control espoused by Colbertism. As Boissonnade observed, the principal objection to Colbert's system was that "it introduced inflexible bureaucratic methods in a domain where individual initiative must have a principal role. The system did not take into account regional or local aptitudes."[29]

---

[25]The number of royal inspectors never matched the thousands of U.S. army inspectors employed in the 1940s and 1950s. Nor did it match the hundreds of ISO 9000 auditors now employed by more than 50 registrars. Bureaucracy was still in its infancy. Indeed, no more than 20 to 30 royal inspectors operated during the 17th and 18th centuries.

[26]Martin, *La Grande Industrie*, p. 123. "Un inspecteur local est chargé de veiller a toutes les oper-ations de chaque fabrique; de visiter definitivement les draps après les derniers apprets et de les marquer ou les rejetter. Le seul effet de ces inspections se reduit aux frais de bureau, à la degrada-tion des marchandises, au retardement des expeditions, à l'oppression des fabricans qui n'osent meme pas se plaindre des injustices des inspecteurs, parce qu'ils seroient exposés a des jugements encore plus rigoureux."

[27]Braudel, *The Wheels of Commerce*, p. 318.

[28]Pierre Dockés, *L'espace dans la Pensée Economique du XVie au XVIIIie Siécle*, Flammarion, Paris, 1969, especially Part I, Les Mercantilistes, pp. 75–77.

[29]Boissonnade, *Colbert: Le Triomphe de L'Etatisme*, p. 53. Il (système Colbert) introduisait des methodes bureaucratiques sans souplesses dans un domaine ou l'initiative individuelle doit avoir le principal role. Il ne tenait aucun compte des aptitudes regionals ou locales.

Is it fair to conclude that the sole heritage of Colbertism was the introduction of a costly and rigid bureaucracy which, as Coles suggests, only helped to retard the development of French capitalism? Historians such as Coles, Boissonnade, Dockés, and others cited above generally agree that the age of Colbertism had mostly a stifling effect on France's 17th-century economy. Still, in view of and perhaps despite these objections, can we then assume that Colbert's system generally helped or hindered the quality of French products? Would the quality of French products have improved at a faster rate without Colbert? That is, of course, a very difficult if not impossible question to answer. One could try to compare British, Italian, or Flemish industries with French industries during the same epoch. But it would be difficult to assess which country produced the best products. Unfortunately, historians and economic historians rarely spend much time writing about the quality of products. Still, some evidence is available. One should recall that when Colbert exhorted French manufacturers to improve the quality of their products, he used as models the manufactories of Flanders and Italy. In today's parlance, one could suggest that Colbert was actually benchmarking the manufactories of other kingdoms. We must therefore assume that, at least during the 16th century, Flemish and Italian products were considered to be of a quality worthy of imitation. Yet, to the best of my knowledge, although Italy and Flanders had many guilds that controlled their own industrial standards (similar in concept to the shop-culture found in 19th-century America), Italy and Flanders did not produce a political leader equivalent to Colbert. They did, however, have a vibrant stock market and a long tradition of capitalism that helped promote commerce.[30]

Nonetheless, one should not necessarily conclude that Colbertism made no positive contributions. There is no doubt that, as Frederick B. Artz points out, "[I]n the case of industry, Colbert wished, by improving the training of skilled workers, to raise the quality of French products to the level of those of Italy and Flanders. . . ."[31] This commitment is perhaps one of Colbert's greatest contributions from which France's modern type of technical education, principally the military engineering schools, was to later develop and influence engineering training throughout the world including the United States.[32]

The poor training of workers was certainly a major problem faced by all entrepreneurs. However, in a style that has been emulated by countless employers since then, French manufacturers solved their immediate problem not necessarily by educating the workforce but, rather, by (1) subdividing assembly processes into simpler single tasks, which in turn allowed them to eventually (2) transform their manufactories into factories. This later transformation was achieved by introducing machinery that was far more productive than workers.

---

[30]For a brief but excellent description of the Italian and Dutch stock market during the 14th, 15th, and 16th centuries, see Braudel, *Colbert: Le Triomphe de L'Etatisme*, pp. 97–114.

[31]Frederik B. Artz, *The Development of Technical Education in France 1500–1850*, The Society for the History of Technology and The M.I.T. Press, Cambridge, MA, 1966, p. 51.

[32]See Artz, *The Development of Technical Education*, and Terry S. Reynolds, "The Engineer in 19th Century America," in Terry S. Reynolds (Ed.), *The Engineer in America: A Historical Anthology from Technology and Culture*, The University of Chicago Press, Chicago, 1991, pp. 7–26.

It is evident that, by the 18th century, manufacturers in France and other European countries had begun to recognize the importance of training, documenting, and formalizing their processes. This age of self-regulation and self-monitoring was but a logical extension of the medieval regulations imposed by the guilds to monitor their trade. We now turn our attention to this topic.

CHAPTER 9

The Quest for Repeatability: The Emergence of Factory Organization and Standardization

The rules and regulations imposed by Colbert on French manufacturers were certainly not limited to France; indeed, similar transformations were happening in England. As early as 1750, the subdivision of work into single tasks was practiced at the Whieldon's pottery. In *Objects of Desire*, Adrian Forty writes that in order to achieve greater reliability in workmanship, the manufacturer Wedgwood relied either on retraining the workers or "dividing the labor into yet more stages, which could be supervised more closely."[1] As processes began to be broken down into smaller and smaller steps that could be carried out by different workers, it became necessary to prepare instructions for the various workmen. In turn, this led to the need to design products. As Forty explains, "[T]he work of designing, or modelling as it was known in the potteries, became a distinct and separate stage in the production of pots."[2] It was not long before potters and textile manufacturers discovered that in order to maintain market share, they had to introduce a great variety of products. As customers constantly clamored for new patterns to suit their individual tastes, the challenge was to create great variety "without increasing the cost of production, and without having to accept irregularities and inconsistencies in workmanship."[3] In response to these challenges, manufacturers began to introduce machines (mostly lathes) designed to "accurately" mass reproduce components at a speed never matched by human hands. Mass customization, characterized by flexible manufacturing and short runs, was certainly well known to 18th-century manufacturers.[4]

---

[1] Adrian Forty, *Objects of Desire: Design and Society since 1750*, Thames & Hudson, London, 1986, pp. 32–33. The emergence of management as an entity whose purpose was to closely supervise the worker was therefore already well established by the mid-1700s, long before Adam Smith published his *Wealth of Nations*.

[2] Ibid., p. 38.

[3] Ibid., p. 38.

[4] Stanley M. Davis in his book *Future Perfect* (Addison-Wesley Publishing Company, Reading, MA, 1987) seems to think that the concept of mass customization only arrived in the 1980s, probably when

Yet, although systems of mass production were beginning to be established in Europe by the 18th century, these systems of production were very different from the age of mass production characterized by Fordism in the early 1920s. The necessary ingredient required to modernize this age of pre-Fordism was still missing, namely, mass consumption or consumerism. In the interim, however, a rather unique form of consumerism did emerge during the later part of the 18th century: military consumerism.

Military Mass Production

Commenting on the influence the military had on the development of specialized and standardized mass production, Lewis Mumford observed more than 60 years ago:

The pressure of military demand not merely hastened factory organization at the beginning: it has remained persistent throughout its entire development. As warfare increased in scope and larger armies were brought into the field, their equipment became a much heavier task. And as their tactics became mechanized, the instruments needed to make their movements precise and well-timed were necessarily reduced to uniformity too. Hence along with factory organization there came standardization on a larger scale than was to be found in any other department of technics except perhaps printing.[5]

By 1765, the need for a uniform, standardized system of production that would allow muskets to be produced with such accuracy as to have truly interchangeable parts was recognized by the French general, Jean-Baptiste Gribeauval. Within 20 years, Mumford notes that a Frenchman by the name of Leblanc was already producing muskets with interchangeable parts.[6] By the 1780s, the U.S. War Department began to express some interest in the Gribeauval system, and in 1813 the first contract calling for the production of 20,000 guns with interchangeable parts was issued.[7]

To achieve the required accuracy that would facilitate greater component uniformity, inspection gauges were first introduced at the Springfield Armory in 1819. These gauges were regularly checked against master gauges.[8] Although interchangeability had not yet been fully achieved, a bureaucratic system

---

he first discovered it. See also Joseph Pine III, *Mass Customization*, Harvard Business Press, Cambridge, MA, 1993.

[5]Lewis Mumford, *Technics and Civilization*, Harcourt, Brace and Company, New York, 1934, p. 90.

[6]Ibid., p. 92.

[7]David A. Hounshell, *The American System of Mass Production: 1800–1932*, The Johns Hopkins University Press, Baltimore, 1984, p. 28. The first contract for 500 hundred pistols was issued to Simeon North in October 1798; however, that contract did not call for interchangeable parts.

[8]Jigs and fixtures were perhaps first invented by Jeremiah Wilkinson of Rhode Island in 1776 for the mass production of nails. See Christy Borth, *Masters of Mass Production*, The Bobbs-Merrill Company, New York, 1945, pp. 23–24.

designed to prevent any deviation from the standard pattern (of gauges) was already well developed and imposed on the armory. The manufacturing of fully interchangeable parts was finally achieved in the United States in 1822 when John Hall, using fixtures and three sets of gauges, each comprising 63 different gauges, was able to produce parts with the required accuracy. As Hounshell observes, the (American) "armory system" of production had arrived.[9]

The armory system of manufacturing brought along a significant innovation in the organization of labor that came to be known as the "inside contract system." A logical consequence of the inside contract system was that it introduced and thus anticipated by some 20 to 25 years, the departmental structure of production that was to occur during the 1850s. Another significant contribution was the importance placed on controlling the vast array of instruments, jigs, and fixtures required to achieve high levels of precision. A British observer of the American system by the name of John Anderson, recognized, during his tour of American industries in the 1850s, the importance of the

> *... hundreds of valuable instruments [jigs and fixtures] and gauges that are employed in* testing the work through all its stages, from the raw materials to the finished gun, *others for holding the pieces whilst undergoing different operations such as marking, drilling, screwing, etc., the object of all being to secure thorough identity in all parts ... it is only by means of a continual and careful application of these instruments that uniformity of work to secure interchanges can be obtained.*[10]

Anderson's observations reveal that by the mid-19th century, the principles of quality control were already being implemented. The importance of receiving and in-process and final inspection and testing was certainly recognized.

The ability to produce reproducible parts, challenging as it was for 18th-century industrialists, did not, however, address the more important issue of reliability. Indeed, although French and American muskets could now be assembled from interchangeable parts, this technological feat must have been of little comfort to the soldier who, facing a charging enemy armed with bayonets, had just realized that his loaded musket would not fire. Would he have the time to find another musket and interchange the defective part?

When, during the 1840s government contracts began to decline (until the Civil War), the lessons learned by the armory system at the Springfield Armory and at Harper's Ferry were soon adopted by private companies such as Colt and Sharp. However, as the privatization of the armory system of production by American businessmen began to take place, their organizations also began to undergo some significant structural metamorphoses.

---

[9]Hounshell, *The American System*, p. 50.

[10]Anderson as quoted by Hounshell, *The American System*, p. 64; emphasis added.

The Managerial Revolution (1840–1880): Regulation from Within

It is naturally impossible to pinpoint with great accuracy the exact year when management first made its appearance in America, but Alfred D. Chandler suggests that, "as late as 1840 there were no middle managers in the United States— that is, there were no managers who supervised the work of other managers and in turn reported to senior executives who themselves were salaried managers. At that time nearly all top managers were owners."[11] As long as the geographic extent of the market was limited to the city or its surroundings, the owner-operated company could adequately satisfy the needs of the community. However, with the coming of the railroad, telegraph, and steamship, geographic distances began to shrink and the traditional notion of the local market was drastically modified. Companies quickly realized that the local market was rapidly being transformed into a national market. The notion of mass production and mass distribution now depended, as Chandler observed, on the speed, volumes, and regularity of movement of goods.

As technological innovations began to reshape the concept of market, businesses had to reorganize themselves in order to survive.

Expansion of volume led to the creation of an administrative office to handle one function in one local area. Growth through geographical dispersion brought the need for a departmental structure and headquarters to administer several field units. The decision to expand into new types of functions called for the building of a central office and a multidepartmental structure, while the developing of new lines of products or continued growth on a national or international scale brought the formation of the multidivisional structure with a general office to administer the different divisions.[12]

To coordinate these new multidepartmental enterprises and to monitor the processes of production and distribution, American entrepreneurs had to find new ways to run their businesses. They had to *innovate* and ended up inventing the manager. By the 1850s and 1860s, as businesses were beginning to mass produce standardized consumer goods, the managers were busy coordinating and organizing the transformation of the means of production.

The managerial revolution that began in the late 1840s or early 1850s certainly played a significant role in allowing industry to rapidly adapt to the needs of the mid-19th-century consumer. Still, as production began to expand, and the pressure to produce quality products at a competitive price grew, many of these newly formed enterprises began to recognize the value of the "armory system" of production.

---

[11] Alfred D. Chandler, Jr., *The Visible Hand: The Managerial Revolution in American Business*, The Belknap Press of Harvard Press, Cambridge, MA, 1977, p. 3.

[12] Alfred D. Chandler, Jr., *Strategy and Structure: Chapters in the History of the Industrial Enterprise*, The M.I.T. Press, Cambridge, MA, 1962, p. 14.

Adoption of the Armory System for Private Production

As various industries began to experiment with the so-called New England armory system of manufacturing (later called the American system), each adopted the system to their needs. For example, the clockmaking industry chose not to duplicate the costly bureaucratic structure that was designed to maintain and control the many precision gauges. One obvious reason for not wanting to incur such costs was that clock manufacturers, such as Eli Terry, for example, were under tremendous pressure to mass produce clocks at very competitive prices (a constraint generally not imposed on military production). Also, the precision demanded by the market did not warrant heavy investments in the control of gauges. Therefore, rather than invest in the costly administration of gauge monitoring, pioneers such as Eli Terry focused their attention on the importance of plant layout and process engineering, two important elements of mass production.[13]

What was adequate for the manufacturing of clocks was, however, not necessarily acceptable to other industrialists. Singer's management, for example, recognized by the early 1880s that the so-called European method of manufacturing sewing machines, with general machine tools and much handwork, was not as effective as the armory system of special tools, jigs, and fixtures.[14] New methods were introduced to raise and maintain the quality of the product. Some of these new methods included lot sampling for needles and testing of each steel delivery. In addition, all process-related problems had to be documented, minutes had to be kept of the weekly meetings with contractors, and a gauging department was created. Finally, no changes of model or gauge points was allowed "except in committee of the whole, where a complete record should be kept."[15]

By the 1890s quality control (i.e., inspection) was practiced extensively by some bicycling manufacturers. In some cases, as with Pope bicycles, the system of quality control could be characterized as rigid as the following account illustrates:

> *An inspection department was set up, as in many of the New England armories, with a separate corps of inspectors. Before machining, each drop-forging was inspected. About 5 percent were rejected. The Pope company claimed that its cranks were inspected eight times before being sold and some parts as many as a dozen times.*[16]

---

[13]Borth, *Masters of Mass Production*, p. 29. Terry was therefore able to increase production to several thousands of wooden clocks per year. A set of gauges (ca. 1838) used in the manufacturing of wooden clocks is reproduced in Hounshell, *The American System*, Figure 1.11, p. 55.

[14]The many reproductions of paintings found in Hounsell and other similar works, for example, often show workers filling parts; the "hidden factory" has always been in full production. The days of lean manufacturing had not yet arrived.

[15]Hounshell, *The American System*, pp. 116–121.

[16]Ibid., pp. 207–208. Hounshell also quoted a study by an economic historian which suggests that the British cycle industry did not become competitive until it dropped this rigid system of quality control. It is not clear whether these quality control functions were performed by a department.

All that testing (100 percent of chains were tested) and inspection was not performed without cost. As others entered the market, Pope found it increasingly difficult to remain competitive. Eventually, although Pope remained competitive, it could not maintain its position as the largest producer of bicycles in America.

To ensure the quality of their products, 19th-century American enterprises relied on varied means. McCormick, for example, unable to obtain quality parts from his subcontractors, opted to centralize production, and began, in the 1880s to rely on mechanized production. Others, such as Colt, favored the inside subcontractor system. This ancient system, already practiced by the Venetians in the 14th century and rediscovered at the Springfield Armory, allowed mechanics such as Pratt, Whitney, Spencer, and Fairfield to contract with Colt "to use his shop space, power, machine tools, and materials to produce a particular part or to manage a particular operation for a set piece rate."[17] The practice of inside contracting, as currently practiced by some companies such as Microsoft, for example, was therefore common in the 1860s through 1880s. As Chandler explains in the case of steel and iron mills:

In those departments requiring the most intricate processing techniques in grinding, polishing, and other finishing of metal components, the foremen were responsible for the profitability as well as the productivity of their departments. They frequently became "inside contractors."[18]

Although the practices of inside contracting no doubt encouraged foremen to increase productivity and thus profitability, the question remains as to what impact such practices might have had on product quality. A review of some accounts indicates that, at least in the late 1870s and early 1880s, managers and owners were well aware of the basic principles of quality assurance. Commenting on Mr. Henry Towne's paper ("The Engineer as an Economist") presented at the American Society of Mechanical Engineers in 1885, Mr. W. E. Partridge observed:

(Workmen) can feel an intelligent interest in the success of the establishment. Under a system like that just outlined the loafer is either driven to work or is fired out of the shop. Every man becomes an overseer with powers to act, which he does not fail to exercise, and he does not omit to keep up both standards of quality and quantity. He is a most vitally interested party, becoming virtually a silent partner upon whose co-operation implicit confidence may be placed.

The question is sometimes asked in this connection, How shall we keep up the standard of work as regard to quality? Where machines are built, after the usual inspection is provided for, it is sufficient to guarantee the quality to the customer, and then make the erecting gang responsible for all defects in quality

---

[17]Ibid., p. 49. See also David F. Noble, *Forces of Production*, Oxford University Press, New York, 1986, p. 139.

[18]Chandler, *The Visible Hand*, p. 271.

which they could have detected while erecting. A discoverable error in work-manship, or flaw in material, is charged to those who put the machine together. When those who erect find flaws or poor work, the last man who expended labor on the piece, and who had an opportunity to detect the trouble, sustains the loss. Every man who passes work forward becomes responsible for its quality to the man who next handles the work.[19]

Mr. Partridge's comments, as well as the series of articles which are reproduced in Alfred D. Chandler's *Pioneers in Modern Factory Management*, clearly indicate that as early as 1875 and certainly by 1885, managers had discovered the principles of worker empowerment, internal customer–supplier relationships, in-process inspection, process documentation, and centralized record keeping.[20] However, this does not mean that most managers practiced what they preached.

As America entered the 1870s, industries were about to experience a new challenge, that of trying to achieve profitability amidst recessions.

Controlling the Means of Production Prior to World War I: The Age of Taylorism

The obsession with efficiency popularized by Frederick W. Taylor's scientific management can be attributed to at least two major events: the economic recessions of the 1870s and 1880s and the new fascination with time brought about by the railroads' introduction of standardized time. When, in 1883, the railroads decided to rationalize the 53 time zones that covered the United States, resistance was fierce. What is remarkable about standard time is that railroad managers were able to get the system adopted, even though it had little support and there was no popular demand for it. Indeed, even though states and cities were to resist railroad standard time until at least 1915, Michael O'Malley observes that "[W]ithout benefit of federal law or public demand they (railroads) managed to rearrange the nation's system of public time keeping at a stroke, an administrative and public relations coup of impressive proportions."[21] (*Note:* A similar

---

[19]Mr. Partridge's comments are found in Alfred D. Chandler (Ed.), *Pioneers in Modern Factory Management*, Arno Press, New York, 1979, p. 470; emphasis added.

[20]Frederick W. Taylor, commenting on a paper delivered by Captain Metcalf, already demonstrated his apparent contempt for workers and his fascination with centralized control when he stated: "We have found that any record which passes through the average workman's hands, and which he holds for any length of time, is destined either to be soiled or torn. We have, therefore, adopted the system of having our orders (he was referring to the Midvale Steel Co.) sent from the central office to the small offices in the various departments of the works, in each of which there is a clerk who takes charge of all orders received . . ." (Taylor in Chandler, *Pioneers in Modern Factory Management*, p. 475). Mr. Taiichi Ohno, who was responsible for many of the mass production innovations at Toyota during the 1940s and 1950s, would no doubt have agreed with Mr. Partridge's comments. For a summary of Mr. Ohno's accomplishment, see James P. Womack, Daniel T. Jones, and Daniel Roos, *The Machine that Changed the World*, Rawson Associates, New York, 1990.

[21]Michael O'Malley, *Keeping Watch: A History of American Time*, Penguin Books, New York, 1990, p. 100.

coup was to be achieved with the marketing of the ISO 9000 standards, to be explored in Part IV.)

By the mid-1880s, advertising agencies were doing their best to convince the public that keeping time or "keeping watch" (and thus punctuality) was a virtue since it was associated with discipline and vigilance. Before long management learned to use the clock as an instrument of regulation, control, and surveillance of the industrial workforce. The economic pressures of the recession coupled with management's infatuation with efficiency, discipline, and punctuality no doubt influenced Taylor's "discovery" of scientific management and efficiency engineering.[22] One of the consequences of Taylorism is that "it systematized and standardized the processes of production in order to concentrate control over them in the hands of management."[23] As we shall see, this was to have a profound impact on the standardization movement.

Although Taylorism was initially applauded by some industrialists, notably William Sanders who had hired Taylor to perform his efficiency studies at Midvale, signs of rebellion, reminiscent of the resistance to Colbert's regulations, soon followed. Some of Taylor's earlier admirers such as Henry Gantt and Frank Gilbreth began to defect. The quest for efficiency took on a different path. In his attempts to humanize management, Gantt emphasized the "task and bonus system," which rewarded workers for meeting quotas. Similarly, rather than analyzing the timing of motion, which did not take into account human variability, Gilbreth now focused his research on trying to capture the best motion. Taking advantage of the new technology of motion picture and time-lapse photography, Gilbreth began to produce "chronocyclographs" or light path motion models. Encouraged by his efforts, Gilbreth then proceeded to find ways to graphically break down and efficiently document processes and tasks. His efforts led him to the "Therblig," a rather clumsy form of flowcharts that was never adopted.[24] By the turn of the century, psychologists such as Lillian Gilberth and Hugo Munsterberger were beginning to investigate the effect of fatigue (promoted in part by the rigor of Taylorism) on efficiency.

Yet, even though Taylorism was challenged by many, its principles were independently reinvented by Ford and his army of engineers. Despite the similarities between Fordism and Taylorism, there was one important difference. Whereas Taylorism constantly attempted to improve worker efficiency by optimizing their operations, Fordism achieved ever greater efficiency and speed by introducing more machinery.

---

[22]Ibid., Chap. IV, "Keep a Watch on Everybody." As with most movements, Taylorism and its emphasis on time motion studies, was not really new. Gideon writes that time studies designed to compute the cost of each operation were performed by Peronnet in 1760 and Charles Babbage in 1832. See Siegfried Gideon, *Mechanization Takes Command*, Oxford University Press, New York, 1948, p. 114. Taylor did, however, produce a complete "theory" of scientific management.

[23]David F. Noble, *America by Design*, Oxford University Press, New York, 1977, p. 82.

[24]Pictures of chronocyclographs are reproduced in Gideon, *Mechanization Takes Command*, pp. 28, 102–104. An example of a "Therblig" (Gilberth spelled backwards almost) is found in O'Malley, *Keeping Watch*, p. 236.

Controlling the Means of Production: The Interwar Years (1915–1939) and Company Standardization

As early as 1908–1910, engineers at Ford were continuously improving the efficiency, flow, and speed of the production line by introducing special-purpose machinery. As engineers began to introduce an ever increasing number of machines, they were faced with the same challenge, experienced some 150 years earlier by 18th-century manufacturers: How should workers be trained to operate the machinery? Part of the solution was to design fixtures that would be easy to handle (hence, require less training) and still maintain the required accuracy. Next they realized that to maintain repeatability, they had to standardize procedures. To address that issue, Ford engineers wrote guides known as operation sheets for every subassembly and assembly (much as management had done at Singer some 30 years prior). Although these operation sheets contained instructions that "detailed the machining operations on various parts, the requisite material inputs, and the necessary tools, fixtures, and gauges (all of which were numbered and referenced to drawings of parts,)" they were never conceived to be rigid instructions.[25]

By 1908, Ford production engineers had recognized the importance of controlling engineering drawings as well as work instructions, and an elaborate set of document control procedures designed to control and monitor engineering and process changes was introduced. As processes were improved, operations sheets were continually updated. With Fordism, the age of company-wide standardization, which had begun at Springfield and Harper Ferry, was mastered.

The Emergence of the Government as Customer

Despite the success of Ford and other auto giants, signs of an economic slowdown were already present. During the depression of 1929–1936, the industrial capacity of the nation was significantly reduced and companies such as GM and General Electric ran at only 24 percent of capacity. Fortunately, for some companies, notably aircraft manufacturers, "government orders constituted at least two-thirds of the manufacture's business from the mid-1920s until just before the Second World War."[26] It is during that period that the role of the federal government as purchaser of last resort began to be defined. Within a half dozen years,

---

[25]Hounshell, *The American System*, p. 224. Ford had already introduced a type of just-in-time manufacturing system at his Highland Park plant, "where laborers distributed the necessary parts at each station and timed their deliveries so that they reached the station shortly before the parts were needed" (Hounshell, *The American System*, p. 236). Taiichi Ohno, of Toyota, who had visited Ford plants after 1945, rediscovered the system known today as *kanban*. Many industrialists "benchmarked" Ford plants prior to World War I; see Womack et al., *The Machine that Changed the World*, Chap. 3.

[26]Ann Markusen and Joel Yudken, *Dismantling the Cold War Economy*, Basic Books, New York, 1992, p. 41. See also Chandler, *The Visible Hand*, p. 497, and Jay M. Gould, *The Technical Elite*, Augustus M. Kelley, New York, 1966, pp. 22–23.

the importance of the government as a dominant customer was to become clearly established.

The Role and Influence of the Military as Customer of Last Resort

As the United States entered World War II, a new era of customer–subcontractor relationships was about to be created. Prior to the 1930s, and even during the 1940s, supplier–customer relationships were not dictated by any rigid regulations. Certainly, over the centuries, customers have never missed an opportunity to inform their subcontractors of inadequate workmanship. During the 19th century, the American mechanician and industrialist Nathan Sellers, who subcontracted to several shops the manufacturing of the first fire engine, was keenly aware of quality. David Hawke tells us that "[I]f the quality of the work failed to satisfy Sellers, the subcontractors heard from him."[27] During the age of Colbertism (Chapter 8), we also saw how the king's auditors were responsible for auditing manufacturers for the quality of their products.

Yet, despite a history of regulation, the age of supplier evaluation characterized by second-party (customer) audits and the obligatory lengthy and often irrelevant questionnaires of recent years had not yet arrived. This was partly due to the fact that major manufacturers, such as Ford, subcontracted very little of their work. The Ford plants at Rouge and Highland Park, for example, were so vertically integrated, complete with steel mill, glass factory, and metal-forming and metal-cutting departments that little, if any, work was subcontracted. This tradition lasted until the 1950s. In the case of General Motors, the Ford approach was modified in the 1920s by Alfred Sloan—who had also introduced yearly model changes and flexible manufacturing—when he set up decentralized parts-making divisions as independent profit centers.

As the United States was readying itself to enter World War II, it became clear that the private sector had to rapidly convert to war production. The transition, described in Christy Borth's *Masters of Mass Production*, was rapidly achieved not so much thanks to American ingenuity as Borth would have us believe but rather, because American production was, as we have already noted, operating at less than 50 percent capacity. As the Department of Defense became the sole customer, it began to impose on its suppliers specific quality requirements. These requirements had an impact on American and world industry that continues to this day.

---

[27]David Freeman Hawke, *Nuts and Bolts of the Past: A History of American Technology 1776–1860*, Harper & Row, New York, 1988, p. 37.

CHAPTER 10

Military as Customer and Controller of Subcontractors

Within the U.S. military, the philosophy of controlling the manufacturing processes of subcontractors did not occur until the 1948–1950 period. Prior to that, the military relied on acceptance sampling, which had been implemented during the 1930s and continued throughout the 1940s.[1] The acceptance sampling method essentially relied on taking random samples from lots and inspecting products for nonconformance. Having agreed on a mutually acceptable definition of acceptable quality levels (AQL), the customer then decided whether a lot should be accepted or rejected by referring to a set of statistical tables known as Military Specifications 105 (or Mil Specs). It should be noted that contractors quickly found ways to improve the odds in their favor by simply adjusting the lot size to their advantage.

By the end of World War II, the armed forces had recognized the limitations of acceptance sampling and were looking for a new methodology. Although the Navy Bureau of Aeronautics had developed in 1948 a general document known as T1-48, which outlined basic principles of quality control, the new philosophy was not adopted until the Air Force produced, on December 1950, a new document known as MIL-Q-5923.

Origins of MIL-Q-5923

The following account is based on a series of articles that appeared in 1950, 1954, 1958, 1959, and 1967.[2] In an article that appeared more than 40 years ago, Lt. Col. Griffith of Norton Air Force Base in San Bernardino, California, explains

---

[1]For an early account, see Colonel Leslie E. Simon, "The Advancing Frontier of Quality Control," *Industrial Quality Control*, May 1949, pp. 5–9, and Major General Leslie E. Simon, "Quality Control at Picatinny Arsenal, 1934–1945," *ASQC Technical Conference Transactions*, Chicago, 1971, pp. 273–279.

[2]Lt. Col. O. C. Griffith, "The Air Force Approach to Statistical Quality Control," *Industrial Quality Control*, May 1950, pp. 26–28; Lt. Col. O. C. Griffith, "Air Force Surveillance Inspection and MIL-Q-5923B," *Industrial Quality Control*, September 1954, pp. 9–14; Colonel J. G. Schneider, "What Does the Air Force Expect of Contractors? *Industrial Quality Control*, November 1958,

that following the reorganization of the Air Force in 1947, individuals began to ask how adequate surveillance could be provided within the constraints and limitations of peacetime budgets. In an attempt to address that question, the Air Force began, as early as 1947, and with the assistance of the Stanford Research Institute in 1949, implementing some important changes.

One of the first improvements, initiated in 1947, was the separation of inspection from the procurement function and the renaming of the Inspection Division to "Quality Control." Under the leadership and guidance of such influential individuals as Drs. Grant, Juran, Burr, and others, the Air Force Quality Control program began assigning statistical quality control officers to the new section. The age of statistical quality control had officially arrived; however, this was not sufficient. As Griffith observed: "The Air Force recognized that any kind of 'surveillance inspection' would be ineffective without proper control of its own quality by the contractor."[3] It is this recognition that led to the emergence of the MIL-Q-5923 specification.

Accounts of how the 5923 specification emerged and on how it later influenced MIL-Q-9858 are told by Colonel J. G. Schneider:

"We wrote a specification MIL-Q-5923 (USAF), December 1950, 'Control of Aircraft and Associated Equipment,' which became a part of all major Air Force equipment contracts." Noting that Dr. J. M. Juran "expressed concurrence with our plans and hopes," Schneider proceeds by writing that Specification 5923 "had been translated and in use in Japan and several European countries, and by 1957 more and more people were advocating that it be adopted as a Department of Defense standard. . . . It was my pleasure and honor," Schneider writes, "to represent the Air Force in negotiating a document to be used as a DoD standard quality control system with Army and Navy representatives. The standard adopted was MIL-Q-9858 dated April 8, 1959." It is interesting to note that Schneider wanted to emphasize that "MIL-Q-9858A (the 1961 updated version of 9858), goes much further than the 5923 series and 9858 in that it requires a quality *program* rather than a quality control *system* and it elaborates on basic policies and philosophies and better defines the planning of the elements of the original documents."[4]

---

pp. 12–14; John J. Riordan, "Quality Control Management in the Department of Defense," *Industrial Quality Control*, December 1959, pp. 11–13; Harold L. Kall, "Does Your QC Organization Meet MIL-Q-9858?," *Industrial Quality Control*, January 1963, pp. 10–12; and J. G. Schneider, "Quality Planning—The Key to Pursuit of Perfection," *Industrial Quality Control*, July 1967, pp. 4–6.

[3]O. C. Griffith, "Air Force Surveillance Inspection and MIL-Q-5923B," *Industrial Quality Control*, September 1954, p. 10.

[4]Schneider, "Quality Planning," p. 5. Apparently the word "program" did not yet acquire the pejorative meaning now associated with the word, at least in the United States. Today, we would probably prefer the word "system," as indeed the 9000 series emphasizes. John J. Riordan writes that, prior to the publication of MIL-Q-9858, a series of Department of Defense (DoD) instructions were published. Some of these key instructions were DoD Instruction 4155.6, "Department of Defense Quality Assurance Concept and Policy," April 14, 1954; DoD Instruction 4155.8, "Department of Defense Procurement Inspection Policies and Procedures for Items Covered by Military and Federal Specifications," May 7, 1957; and DoD Instruction 4155.10, "DoD Policies and Procedures for Assuring the

The above accounts reveal some interesting facts: (1) MIL-Q-5923 was first published in December 1950, (2) it is referred to as a "specification," (3) it was written for "The Control of Aircraft and Associated Equipment" (to be expected since this was an Air Force specification), (4) it had been adopted by several European countries and Japan by 1957, and (5) it strongly influenced MIL-Q-9858.

Justification for MIL-Q-5923[5]

It is interesting to review some of the arguments presented more than 45 years ago as to why MIL-Q-5923 was necessary. Besides the usual economic arguments, which suggest various cost savings for the Air Force, I believe it is important to understand and recall the customer–supplier relationship that led to the emergence of 5923. Readers familiar with ISO 9000 (see Chapter 12) may want to compare and contrast the events of 45 years ago with events surrounding the ISO 9000 phenomenon.

Reviewing the Air Force policy as specified in the (1954) Armed Services Procurement Regulations, Griffith observed that "This policy recognizes, first of all, the unique position of the government as a consumer. This is dictated primarily by the fact that payment for the material it buys is provided by public funds. . . . Further, all citizens have the right to expect that the government will get full value for the money it spends for its material." Noting that, "*Government inspection, then, is a legal requirement*," [emphasis added] Griffith goes on to proclaim that such requirements will guarantee that the government (i.e., taxpayers) "will get a fair return for every dollar it spends." Wonderfully optimistic and perhaps even naïve, but this was the 1950s and many more thousands of military standards had yet to be written and the $750 screwdriver had not yet arrived.

Rather than rely on an army of government inspectors, Griffith suggested that "it is certainly far more logical and practical to require that each contractor inspect and test his own product, and then furnish *objective evidence* that it meets all specified requirements." This certainly appears to be a reasonable proposition, but I question the logic of the argument. Indeed, all that the Air Force was doing was shifting the cost of inspection from one source to another. The inspection cost did not disappear because the law still required the inspection to be performed! Certainly, the Air Force could now claim that its army of inspectors had vastly been reduced (from 15,000 during the war to 1100 after the war), but they merely reappeared somewhere else; namely, with the many contractors who now had to adopt Specification 5923. Dollars had to be spent one way or the other, and taxpayers ended up paying the bill anyway

---

Quality of Production of Complex Supplies and Equipment," February 10, 1959. John J. Riordan. "Quality Control Management in the Department of Defense," *Industrial Quality Control*, December 1959, pp. 11–13.

[5]Griffith's quotation for this and other sections are from his 1954 article, "Air Force Surveillance," pp. 10 et passim.

since the cost of inspection was no doubt passed on by the contractors to the Air Force. Did the nation really save money? Probably not, and Lt. Col. Griffith does seem to indirectly acknowledge that fact when he writes a few lines later: "... the government buys not the physical article, but a service, the service of a contractor's inspection to assure conformance to the article to contractual requirements."[6]

Early Resistance to MIL-Q-5923

As early as 1952, J. H. Sidebottom of the Aircraft Industries Association of America, Inc., voiced some legitimate complaints regarding 5923:

Our idea of the most productive arrangement between the government and aircraft industry is one in which we should operate as closely as possible to the strict buyer–seller relationship. This implies that emphasis be placed on the end results rather than on the method or procedure. The goal of policies and regulations should be to establish broad objectives, leaving the procurement agencies and the industry the freedom and flexibility they need to negotiate the technical terms and prices to do the job in the best way they know how.[7]

Sidebottom objected to the indefinite wording of the specification which led to a "multiplicity of interpretations," particularly those that were rendered a few steps "down the ladder." Sidebottom offered numerous pertinent observations and recommendations:

- The military should unify and simplify "their ponderous paper work and traditional form filing procedures."
- The people who write the directives and specifications should get out to see how well they are working in the field.
- The inspectors are called on to perform functions for which they are not qualified.
- The responsibility for quality control should be vested in the organization "which is manufacturing the article or product."
- The current (1952) system is not efficient and economical.[8]

---

[6]I believe it is worth noting that although the ISO 9000 series is not sponsored by one powerful customer, equivalent to the Air Force, a similar series of economic and noneconomic events is currently evolving worldwide. The Department of Defense has only recently officially recognized the ISO 9000 series but does not require its suppliers to be ISO 9000 certified. I suppose the argument could be made that the European Community could be perceived as the nearest equivalent to the Air Force in 1954. However, the European Community is still far from representing one united front. We are thus left with the obvious question: What organization(s), if any, has taken on the role of the Air Force and why?

[7]J. H. Sidebottom, "Military Inspection in the Aeronautical Field," *ASQC Quality Conference Papers*, 1952, p. 41.

[8]Ibid., various pages.

Although Sidebottom's observations have since proven to be true, as witnessed by recent statements made by the Department of Defense, his is one of the very rare incidences of constructive criticism ever published within the pages of any of the ASQC's transactions or *Industrial Quality Control.*[9] Why is that so? Several arguments could be proposed. Had they read Sidebottom's article most military officials would have no doubt proposed, as Griffith did in 1950, that the argument was futile simply because of clause 5e of the Standard Contract Form Number 32. The clause clearly specified that "The contractor shall provide and maintain an inspection system acceptable to the government covering the supplies here under. Records of all inspection work by the contractor shall be kept complete and available to the government during the performance of this contract and for such longer periods as may be specified elsewhere in this contract."[10] The rules of the game were, therefore, rather simple: If subcontractors wished to have the Department of Defense as a customer, they had to accept the law.

The Omnipresent Customer

The quality assurance system imposed by the Department of Defense during the past four decades has been burdensome to America's leading contractors and subcontractors. Yet, among the chorus of praise and patriotic arguments propagating the belief that the system of rules and regulations contained within the volumes of the Armed Services Procurement Regulations (ASPR) were logical and justified, a few dissenting voices could be heard. During the past 45 years, one leading exponent of the deleterious effect of so-called "Pentagon management" has been Seymour Melman. Commenting on the peculiar buyer–seller relationship found between the Department of Defense and a contractor, Melman observed:

> ... *the overriding fact is that the selling firms sell to one customer. Moreover, this customer is not replaceable, that is, there is no diverse market of other customers—other Department of Defense—to whom the supplier firms might sell their products. Apart from the existence of one customer and the obvious leverage that this gives to him, there is also the fact that this "customer" has organized himself for controlling, in depth, the affairs of the supplying firms.* ...

---

[9]Among the few articles somewhat critical of the military specs one could cite Robert L. Cooley, "Are Mil Specs Obsolete?," *Annual Technical Conference Transactions*, 1969, pp. 29–32, in which the author suggests that military specifications must be upgraded with the help of industry. I am referring to the June 29, 1994, memo issued by Secretary of Defense William Perry and titled "Specifications & Standards—A New Way of Doing Business."

[10]A. F. Cassevant, "Army Quality Control and Inspection," *ASQC National Convention Transactions*, 1957, p. 507. See also William K. Ghormley, "Contract Requirements and Quality Control," *ASQC National Convention Transactions*, 1957, pp. 511–516, where the author concludes by stating, "I hope that you will find that the procurement quality kit-of-tool is adequate to permit both industry and government to perform their respective role in the shaping of a stronger America" (p. 516). Could anyone challenge such a puerile statement without running the risk of being perceived as antipatriotic?

> *As a result of these controls, the relation between buyer and seller becomes a relationship between top management and subsidiary management.*[11]

By the late 1960s, as many as 9000 people were employed by the Defense Contract Administration Services to check military contractors for compliance with a host of military procurement specifications and other regulations. Was the quality of products improved thanks to the bureaucracy? Did military contractors ever consistently achieve the elusive goal of zero defects and, if so, at what cost? The evidence provided by Melman and others would suggest that the huge bureaucracy that grew out of the Defense Contract Administration Services did not necessarily enhance the quality and reliability of the product. These products were in fact so complex that the percentage of "Not Mission Capable" aircraft for the Air Force and the Navy averaged 42 and 38 percent, respectively.[12] Military contractors had learned to, in the words of Melman, profit without production.

This ability on the part of the military economy to maximize cost had been acknowledged at least as early as 1974. In a paper presented at the 1974 ASQC Technical Conference in Boston, Colonel Theodore G. Zeh openly admitted that historically, Department of Defense quality assurance personnel had been concerned principally with assuring that a quality product was accepted by the government. "Cost," Colonel Zeh observed, had "been secondary." "As a result of the quality assurance representative's preoccupation with quality," Colonel Zeh pointed out that it was not unusual for quality people "to allow the contractor to incur additional cost in assuring a given level of quality, even if the contractor by so doing imposed himself an inspection plan much tighter than the actual government requirements."[13]

There were, however, a host of other reasons as to why the cost of production associated with military contracts continued to rise. The desire to monitor almost every aspect of a contractor's operations has certainly been a major contributor. Within the aircraft industry, for example, David Noble notes that "management

---

[11]Seymour Melman, *Pentagon Capitalism*, McGraw-Hill Book Company, New York, 1970, pp. 36–37. Jacques Ellul observed the same point when describing Colbert's administration (Jacques Ellul, *Histoire de Institutions de l'Epoque Franque à la Revolution*, Presses Universitaires de France, Paris, 1967, pp. 397–438).

[12]Seymour Melman, *Profits without Production*, Alfred A. Knopf, New York, 1983, p. 217. See also his Chap. 8, "Opulence in the State Economy" for an analysis of the profits without production theme. Such statistics certainly did not agree with the zero-defect (ZD) movement that emerged in the early 1960s. This movement proposed that zero defects could be achieved if only people were motivated enough to achieve perfection. Lack of knowledge or lack of attention, Philip Crosby proclaimed in 1964, were the real sources of mistakes. See Philip B. Crosby, "Z is for Zero Defects," *Industrial Quality Control*, October 1964, pp. 182–185. For a brilliant criticism of the ZD movement, see J. M. Juran, "Quality Problems, Remedies and Nostrums," *Industrial Quality Control*, June 1966, pp. 647–653.

[13]Colonel Theodore G. Zeh, "DOD—At Home and Abroad," *ASQC Technical Conference Transactions*, 1974, pp. 449–450. This maximization of cost was later more fully documented by Melman and J. R. Fox. See, for example, Appendix III, "How the Military Economy Maximizes Cost," in Melman, *Profits without Production*, pp. 301–304.

had nearly doubled as well, in an effort to insure control over production and thereby guarantee performance according to military specifications."[14]

These increases in managerial control led to other forms of process control. Indeed, as anyone who has worked within an organization dependent on military contracts knows, government auditors do not usually tolerate undocumented changes, no matter how insignificant. This has led to some frustrating and costly anecdotes. Noble recounts the story of an operator who could not make any changes, no matter how minor, to his numerically controlled machine, without prior approval from the Air Force. This bureaucratic red tape imposed by the Air Force and other government agencies became so complex that contractors were/are forced to constantly find creative, but often expensive, work-arounds. Thus, rather than change a tool, which requires customer approval and results in delays, one simply adjusts a fixture instead.[15]

Reflecting on the significance and influence of the American system of production on American industries, Noble concluded:

> *Rather than any such economic stimulus, the overriding impulse behind the development of the American system of manufacture was military; the principal promoter of the new methods was not the self-adjusting market but the extra-market U.S. Army Ordnance Department. The development of interchangeable-parts manufacture began within the arsenals as an "expensive hobby of a customer with unlimited funds and was dictated by military criteria of uniformity and performance, regardless of cost." It was subsequently encouraged and carried over into civilian production (agricultural implements, sewing machines, bicycles) by arsenal personnel who brought with them a military enthusiasm for uniformity and automation that reinforced the growing industrial obsession, epitomized by Andrew Ure, with "perfecting" production by eliminating labor.[16]*

The transfer of personnel from the Arsenal to civilian production continued for the next 100 years and is likely to continue and even increase as all Armed Forces continue to reduce their staff. Over the years, I have found throughout the course of my visits to plants across the United States, that many quality managers once worked for a defense contractor. These individuals usually end up implementing at their new work site the quality assurance of their former employer.

---

[14]David E. Noble, *Forces of Production*, Oxford University Press, New York, 1984, p. 6.

[15]The story of how the Air Force influenced the development of N/C technology and how this influence eventually affected the competitiveness of American tool manufacturers is carefully documented by Noble (ibid., p. 85): "The Air Force," Noble observes, "in its development of high-performance fighter aircraft, was confronted with unprecedented machining requirements. The complex structural members of the new aircraft had to be fabricated to close dimensional tolerances and this extremely difficult and costly process seemed to defy traditional machining methods. . . . The result was N/C technology, which also happened to meet military requirements for greater control over production (for quality control and "security" purposes) and manufacturing flexibility. . . ."

[16]Ibid., p. 334.

By the mid-1960s, a time when nearly 40 to 45 percent of all articles published by *Industrial Quality Control* were either written by members of the Armed Forces or by authors working for contractors to the Department of Defense, most everyone believed that a supplier surveillance program as developed and enforced by the government was indeed the only way to ensure quality. How was this consensus achieved? Why and how was the ponderous, bureaucratic, and costly military system of inspection adopted by the civilian sector? Finally, why did the International Organization for Standardization choose to recycle, and in essence privatize, these ponderous and occasionally anachronistic military inspection dictates into the ISO 9000 series? We will now attempt to answer these and other questions.

IN THE BEGINNING

God created heaven and the earth. Quickly he was faced with a class action suit for failure to file an environmental impact statement. He was granted a temporary permit for the project, but was stymied with the cease-and-desist order for the earthly part.

Appearing at the hearing, God was asked why he began his earthly project in the first place. He replied that he just liked to be creative.

Then God said, "Let there be light," and immediately the officials demanded to know how the light would be made. Would there be strip mining? What about thermal pollution? God explained that the light would come from a huge ball of fire. God was granted provisional permission to make light, assuming that no smoke would result from the ball of fire; that he would obtain a building permit; and to conserve energy, that he would have the light out half the time. God agreed and said he would call the light "Day" and the darkness "Night." Officials replied that they were not interested in semantics.

God said, "Let the earth bring forth green herb and such as many seed." The EPA agreed so long as native seed was used. Then God said, "Let waters bring forth creeping creatures having life; and the fowl that may fly over the earth." Officials pointed out this would require approval from the Department of Game coordinated with the Heavenly Wildlife Federation and the Audubongelic Society.

Everything was OK until God said he wanted to complete the project in 6 days. Officials said it would take at least 200 days to review the application and impact statement. After that there would be a public hearing. Then there would be 10 to 12 months before. . . .

At this point God created Hell.

Source: Story provided by a Spanish Internet pen pal.

Part IV

The Age of Standardization

As Colbert discovered long ago, one of the unfortunate side effects of regulations is that, once written, a bureaucracy must be created to enforce the regulations. In the case of the ISO 9000 series of quality system standards, the bureaucracy consists of the numerous committees that are busily writing more guidelines and redesigning standards, as well as registrars and the army of auditors they employ to monitor companies.

Although these international standards were always intended to be voluntary, the persistent *de facto* myth, fortunately not yet universally accepted by all, is that if one wants to please customers or gain market share, ISO 9000 registration must be achieved. The problem with this assumption is that as registration requirements are routinely passed down the hierarchy of suppliers, the burden of implementation becomes proportionally greater. When one takes into consideration the already heavy regulatory demands placed on most businesses by the many layers of city, county, state, and federal regulations and bureaucracies and varied customer demands, the requirements for one more international standard become a difficult pill to swallow for most small businesses. In the state of Washington, for example, a potato chip manufacturer with 40 employees is regulated by as many as 16 agencies ranging from the Washington Department of Agriculture to the Washington Safety and Health Administration. A truck repair company with 10 employees is regulated by 11 state agencies.[1]

The problem with bureaucracies is that they cannot adapt rapidly enough to the ever increasing rhythm of changes demanded by the age of faxes, cellular phones, electronic mail, and the so-called Information Superhighway. But rapid response to change is precisely what organizations must now be able to achieve in the 1990s. Michel Crozier, anticipating Alvin Toffler and quality expert Tom Peters by more than 20 years, proposed as early as 1963 that "no modern organization can survive unless it remains sufficiently flexible and adaptable."[2] To

---

[1]Richard Buck, "Weight of the Law," *Seattle Times*, March 20, 1994, pp. D1, D4. In some cases federal regulations can lead to amusing anecdotes as the following newspaper heading reveals: "Federal laws causing toilet shortage." *Daily Journal of Commerce*, Seattle, July 13, 1994, p. 5.

[2]Michel Crozier, *Le Phenomene Bureaucratique*, Presses Universitaires de France, Paris, 1963, p. 228.

achieve that goal, Crozier noted that the modern organization must trust certain groups or individuals with their initiative and ability to invent. Innovation and the ability to invent is not what international standardization committees are good at.

Part IV explores how the standardization movement developed during the past 50 years. The privatization of military standards and the bureaucratic impact of standardization as manifested in the ISO 9000 series of standards are also analyzed.

CHAPTER 11

The Value of Standardization: Point Counterpoint

Introduction

As I began to write this chapter, the screen on my computer monitor began to periodically turn green. After a few days of tapping and then hitting the side of the monitor, I had to face the fact that it was time to replace my 18-month-old monitor! As I unpacked my new "Made in South Korea" monitor I could not help but notice the long string of acronyms listed on the bottom of the monitor's box as well as in the back of the monitor. The monitor had *six* seals of approval: two from Germany (GS and TUV), one from the United States (UL), one for Canada (CSA), one for Sweden (TCO), and one for the European Community (CE mark). In addition the monitor satisfies FCC, DHHS, and MPRII specifications. Surely this monitor would last more than 18 months (I hoped), but were all of these seals of approval what international standardization was all about? Individual certification for each country?

Encouraged by the success of the ISO 9000 series of standards, and aware of the economic and marketing rewards of having a national standard adopted by the international community, some countries have been developing their own sets of standards which they hope will one day be adopted by the international community. (Sweden's TCO, which includes ergonomic as well as health and safety requirements, is one such example.) The British Standards Institution (BSi), for example, reaped enormous profits and prestige throughout the world when its BS 5750 standard—which itself was derived from U.S. and NATO military standards—was adopted in 1987 by the International Organization for Standardization as the ISO 9000 series. Its *Occupational Health and Safety Management Systems—Specification* (OHSAS 18001-1999) may achieve similar success in Latin America.

If, as is likely, European countries continue to develop more standards to address the ever increasing number of European directives generated by the insatiable bureaucracy in Brussels (site of the European Parliament), then one can be sure that within the next 3 to 5 years the same computer monitor will come with a 5- to 10-page booklet listing all of the electronic, electromagnetic, environmental, and ergonomic standards available. I wonder how many certifications the South Korean manufacturer had to obtain 10 or 15 years ago? Simplicity, where have you gone?

The Value of Standards

The value of standards, especially product standards, has long been recognized by countless authors. Dickson Reck summarizes it best when he writes that "[I]ndustrial standards are concerned with the *uniform*, the *widely accepted*, and the *best* for the time being, even though that best may be only a 'practical-best,' arrived at as the product of a compromise of interested parties."[1] Reck's reference to "interested parties" is worth remembering, for as we shall see in a subsequent chapter, the push for certain types of standards, product or otherwise, can lead to substantial—and often unanticipated—economic rewards for some organizations.

Of the many cases that have been cited to demonstrate the broad range of benefits brought about by product standardization, a difference in screw threads is perhaps the most often cited example. The material and socioeconomic costs associated with such lack of standardization was clearly demonstrated during the Chicago fire of 1871 when fire engines from other cities, which had come to provide assistance, could not connect to the fire hydrants because of a difference in screw threads. After repeated incidents (Boston, 1872; Baltimore, 1904; and San Francisco, 1906), a standard was finally proposed in the 1920s. As a result of the newly adopted standard, lower fire insurance rates were granted to any community that adopted the standardized screw thread. It was not long before the standard became widely accepted.[2]

By the late 19th century, two schools of thought had emerged with regard to the purpose and reasons for standardization: the traditional shop culture made up of veteran engineers, and the school culture favored by the emerging class of younger engineers. The ideology that separated these two cultures was directly influenced by the educational training and experience of engineers. The shop culture of mechanical engineers with its homologue, the "field culture" of civil engineering, represented the traditional "school of experience." The school culture represented the college-trained engineers who were beginning to graduate from the newly founded engineering schools. To shop-culture engineers, like William Sellers (creator of the American standard for screw threads), the emphasis on standardization was a necessary prerequisite to improve shop efficiency.[3]

---

[1]Dickson Reck (Ed.), *National Standards in a Modern Economy*, Harper & Brothers, New York, 1954, p. 18.

[2]Earle Buckingham, "Company and National Standards in Interchangeable Parts Manufacture" in Reck (Ed.), *National Standards in a Modern Economy*, p. 96. For a good history of the Bureau of Standards, see Gustavus A. Weber, *The Bureau of Standards: Its History, Activities and Organization*, The Johns Hopkins University Press, Baltimore, 1925.

[3]As president of Midvale Steel, Sellers was very much interested in establishing managerial control over the various shop processes. To achieve that objective, he hired Frederick Taylor who proceeded to rationalize operations as well as systematize and standardize the processes of production. See David F. Noble, *America by Design*, Oxford University Press, New York, 1977, pp. 27, 40, 76–82.

These engineers would work with associations such as the American Society of Mechanical Engineers (founded in 1880) or the American Society for Testing and Materials to help develop and promote interindustry standards. Yet, although they advocated the need for standardized screw threads, gear teeth, bolts, and so on, most high-ranking (shop-culture) managers did not accept standards imposed from outside the industry. This standardization from without the industry was perceived as a form of intrusion and interference that went against the basic principles of private enterprise. Thus, although shop-culture managers acknowledged the value of industry standards, they felt that the market, rather than standards, had to decide who should win. In other words, it was assumed that the manufacturer who dominated a particular market would be entitled *de facto* to determine the standards for that market (much as Microsoft has done with its Windows™ software, for example).

In contrast, the school-culture engineers "advocated some centralized authority, preferably governmental, which would set absolute standards to which all industries must conform. They assumed, of course, that school-trained engineers in government would set those standards."[4] Judging from the rapid and continued growth of regulatory agencies, and the promising, yet still unproved success of international standards for quality assurance such as the popular ISO 9000 models, it would appear that the school culture is certainly thriving and may even one day dominate.[5] Still, the polarization between the two cultures persists to this day and one could propose that the current worldwide attempts at standardizing quality assurance systems, as typified by the ISO 9000 phenomenon, for example, will actually help accentuate the schism—a point to which we shall return in a later chapter.

By the late 1890s and early 1900s, several trade associations and industries, particularly the electrical industry, had developed their own standards. Not to be outdone, the U.S. government began to develop standards for government purchasing. The U.S. Navy Department standardized its purchases in what has been referred to as "one of the most complete, thorough and best articulated sets of supply standards and specifications of any organized body, either governmental or private."[6] By 1928, the Federal Specifications Board had promulgated a mere 550 specifications, which, in retrospect, is rather puny when compared to the more than 31,000 specifications issued by the Department of Defense.[7]

The value and benefits of standardizing processes and operations to mass production was already well recognized by the early 1920s, a period that saw Taylorism continue to prosper as a logical extension of standardization. Indeed,

---

[4]Ibid., p. 76.

[5]As of 1995, it was estimated that the Food and Drug Administration alone employed as many as 9500 to 9600 functionaries; see Tony Snow, "A Government Full of Regulatory Junkies," *USA Today*, August 8, 1991, p. 11A.

[6]*Industrial Standardization*, National Industrial Conference Board, New York, 1929, p. 139. On the same page the author also informs us that "The first recognition of the necessity and value of buying according to standards and specifications came with the establishment of the General Supply Committee by Congress in 1910."

[7]See "DOD Streamlines Its Purchasing Practices," *Quality Progress*, September 1994, pp. 15–16.

one could read in a 1929 book on industrial standardization that once process standardization had been achieved, "the basis has been laid for the application of time and motion principles and the working out of performance standards involved in scientific management."[8] The connection between standardization and scientific principles had been made.

Origins of the International Organization for Standardization

Most observers would acknowledge that the emergence of the international standardization movement during the early 1920s can be directly attributed to several factors, including (1) the continued evolution and growth of mass markets and (2) the mass production and mass distribution system that developed to satisfy the ever increasing demands of mass consumption, which in turn influenced (3) the need to constantly improve rapid manufacturing; and, finally, (4) the growth of international trade that followed World War I.

By 1926, the 14 countries that already had national standardizing bodies decided to establish the International Federation of the National Standardizing Associations (ISA), which operated until 1939.[9] In 1943, an interim organization was set up under the name United Nations Standards Coordinating Committee (UNSCC). With the end of World War II, the need for a permanent international standardization body was once again recognized. During the October 1946 organizational meeting of the UNSCC, held in London, it was decided that the new association

> *. . . should be in the form of a merger of the original Federation (ISA) and the UNSCC, and that the technical work of both organizations should be consolidated and continued. The name finally agreed upon was the International Organization for Standardization, and its symbol—ISO—was created. The first meeting of this newly formed association was held in Zurich in June 1947.*[10]

During the past several decades approximately 13,000 standards have been published by the International Organization for Standardization, including the notorious ISO 9000 family of standards.

Regulations: Who Are the Interested Parties?

A popular argument presented in favor of government regulations or international standardization is that they benefit the public. This is certainly true in many cases; however, one must also remember that few regulations are ever written for altruistic purposes; economic, marketing, or competitive advantages are

---

[8]*Industrial Standardization*, p. 41.

[9]The 14 countries were Austria, Belgium, Canada, Czechoslovakia, France, Germany, Great Britain, Holland, Italy, Japan, Norway, Sweden, Switzerland, and the United States. See Howard Coonley, "The International Standards Movement," in Reck, *National Standards in a Modern Economy*, pp. 37–45.

[10]Ibid., p. 40.

usually the motivation. Government regulation of technical standards has been shown to create competitive advantages for certain industries. The German cutlery industry, for example, is said to have benefited from the Solingen law of 1938, which set rigid standards for the quality of the cutlery industry and the right to use the Solingen name (a German city where the cutlery industry is concentrated). And yet, although German cutlery has certainly maintained an excellent reputation for quality, most American restaurants rely on cheap and often useless knives, from Taiwan or Japan (which are certainly very much cheaper than Solingen's knives)! One therefore wonders what, if any, advantage German cutlers obtained from the Solingen law.

Demanding standards can also favor specialized manufacturing and services. American manufacturers of pollution monitoring or control equipment have, until recently, maintained a strong international position thanks to U.S. environmental laws. Sweden's tough standards for product safety and environmental protection have helped Swedish manufacturers compete and, on occasion, challenge American firms on the world market. And in Japan, the decision by the Ministry of Justice to accept facsimile as legal documents helped Japanese manufacturers of facsimile machines position themselves as world leaders.[11]

In some instances, firms or organizations will even lobby for certain regulations if the rules can bring them benefits. Established companies in slow growth industries have been known to demand stricter standards simply to raise the entry barrier of potential competitors. This was the case when, in the United States, eastern states—where the industrial base is older—lobbied for tighter national pollution limits for electrical utilities and metal-smelters. The effect was prejudicial to younger rivals, which operated in southern and western states. Companies can also lobby for standards they know they can meet but that will impose high costs on the competition (ARCO, for example, with its EC-1 gasoline). Finally, companies will press for regulations that will create a market for their product. This is typical of the waste management industry, which often allies itself with various environmental groups to press the government for more demanding environmental standards knowing that, when the laws are passed, they may reap the benefits.[12]

Yet, despite the general agreement regarding the value of standards, experts still disagreed as to the role played by the standardization movement. A few, such as Dickson Reck, claimed that ". . . standardization is the great converter of innovations to practical use."[13] Others, argued that standardization is in direct con-

[11]Michael Porter, *The Competitive Advantage of Nations*, The Free Press, New York, 1990, pp. 647–652.

[12]See "Regulate Us, Please," *The Economist*, January 8, 1994, p. 69. Firms do try to bypass the expensive fees charges by cleanup crews by simply controlling the amount of pollution at the source. Either way, money must be spent.

[13]Reck, *National Standards in a Modern Economy*, p. x. For a different point of view, see *Industrial Standardization*, p. 9, where the author notes that the rapid changes in invention and entire reorganization of production conflicted with the need to standardize. For a more recent study, see Charles D. Sullivan, *Standards and Standardization: Basic Principles and Applications*, Marcel Dekker, New York, 1983.

flict with the needs to constantly innovate and invent better processes. Finally, although international standards have often been advertised, perhaps optimistically, as international trade promoters designed to counter the "protectionist" tendencies of national or regional standards, the full effect of the current reorganization of the world's economy into continental trade blocks will not be known for a few years.[14]

The Economics of Standardization

Developing standards can be costly. The initial investment cost associated with the development and publication of the standards includes several factors such as initial research and drafting of the document(s), the time and associated travel expenses of the many committee members who must meet periodically in various cities and who devote their time to review and comment on preliminary drafts, and, finally, the efforts required to reach an industry, national, or international consensus. Once a standard is approved, a process that, for international standards, usually takes 2 to 3 years, additional time and effort must be invested to encourage potential users to consider using the standard. In some cases, the use of a standard may require the development of new testing and inspection procedures. Besides these implementation costs, one must also consider the so-called revision costs periodically required to ensure that a standard does not become outdated.[15]

Once a standard is adopted, the return on investment should more than offset the initial costs. I say "should" because few studies have actually been published that quantify the economic benefits of standards. The studies that have been published, which are for very specific high-tech military applications, do, however, indicate some rather impressive benefit-to-cost ratios ranging from 25:1 to 60:1.[16] This opportunity to save money was recognized by supporters of product standards some years ago. Noting that the adoption of product standards will undoubtedly lead to some savings, Gastineau and Kear confidently observed that "We all know there is money to be made with standards."[17] Gastineau and Kear could not possibly have anticipated the irony of their observation. Indeed, as we witness the growth of the many industries that evolved out of the ISO 9000 movement (consulting, lecturing, registration newsletter, publication, training, etc.), and other related standards (such as QS 9000, AS 9000, TE 9000, ISO 14000 series, etc.), most of us involved with one or more of

---

[14]Reck, *National Standards in a Modern Economy*, p. 191. For a description of how the Japanese use standards as a barrier to trade, see Donald J. Lecraw, "Japanese Standards: A Barrier to Trade?" in H. Landis Gabel (Ed.), *Product Standardization and Competitive Strategy*, North-Holland, New York, 1987, pp. 29–47.

[15]R. B. Toth, "The Economics of Standardization—A Pragmatic Approach," in Robert B. Toth (Ed.), *The Economics of Standardization*, Standards Engineering Society, Minneapolis, 1984, p. 15.

[16]See, for example, Charles E. Gastineau and Donald L. Kear, "Don't Cry: Justify," in Toth, *The Economics of Standardization*, Chap. 3. Gastineau's study is very specific to microcircuit MIL-M—38510/101.

[17]Gastineau and Kear, in Toth, *The Economics of Standardization*, p. 71.

these standards will certainly agree that there is indeed money to be made with standards.

Besides the costs associated with the development of standards, one cannot forget the much larger costs imposed on all manufacturers that must implement the standards (e.g., ISO 9000, ISO 14000, QS 9000, TCO, CSA, etc.). Finally, one must also consider the regulation cost (paid for by taxpayers and companies) involved with ensuring that standards are properly implemented and maintained (auditing, for example). In the case of federal regulations, Tony Snow writes that the number of Food and Drug Administration employees has grown from 7156 in 1985 to an estimated 9576 in 1995. EPA staff has grown 350 percent since 1970 to 18,439 and funding has increased 512 percent. Snow concludes that "EPA regulations impose an average of $7.6 million in economic burdens per year for each life they prolong."[18]

The Limits of Standardization

When authors write about the economic benefits of standardization, they invariably are writing about the positive effects brought about by the standardization of products, specifications, and procedures. As W. Edwards Deming explained some years ago, "the convenience of 110 volts and uniform outlets everywhere in the northern hemisphere would be difficult to express in words."[19] Yet, in the same article, Deming concluded that "standards written by the government become regulations, inflexible. A regulation, in contrast to a voluntary standard, does not permit a manufacturer to depart from a standard in order to develop a specialized and useful business. A regulation," Deming concluded, "eliminates choice by the consumer."[20] This is precisely what Lawrence Abbott had stated 30 years earlier when he noted the potential limitations introduced by standardization with respect to product quality:

> Producers must be free to introduce novel, "nonstandard" qualities. *When innovations in quality are possible but not permissible, quality competition is thwarted. Enforced standardization ordinarily destroys or diminishes quality competition. Other restrictive regulations of a comparable nature, such as local building ordinances, may also lead to its impairment. Of course, when alternatives are prohibited which sensible buyers would never knowingly select, it is only a through-the-looking-glass kind of quality competition that is prevented. On the other hand, when controversial qualities are suppressed (a possible by-product of well-meant regulation) or when special interest groups employ regulation as a device to further there own ends, there is likelihood of a real impairment of socially desirable opportunities for choice.*[21]

---

[18]Snow, "A Government Full of Regulatory Junkies," p. 11A.

[19]W. Edwards Deming, "Loss from Failure of U.S. Industry to Have More Voluntary Standards," *Standards Engineering*, May–June, 1985, p. 52.

[20]Ibid., quoting Senator Ralph E. Flanders, "How Big Is an Inch?," *The Atlantic*, January 1951, p. 53.

[21]Lawrence Abbott, *Quality and Competition: An Essay in Economic Theory*, Columbia University Press, New York, 1955, p. 132.

These restrictions, recognized by Abbott and Deming, have also been explored from a very different perspective by the sociologist Jacques Ellul and others. Quoting the work of Antoine Mal, published during the late 1940s, Ellul writes that "standardization means resolving *in advance* all the problems that might possibly impede the functioning of an organization. It is not a matter of leaving it to inspiration, ingenuity, nor even intelligence to find solutions at the moment some difficulty arises; it is rather in some way to anticipate both the difficulty and its resolution. From then on, standardization creates *impersonality*, in the sense that organization relies more on methods and instructions than on individuals."[22] Moreover, although Ellul agrees with the Italian economist Bertolino who suggested that standardization brought about well-being to the poor, he counters by suggesting that standardization brings about a type of dictatorship. (David Noble proposes a similar argument when he explains how management control over the means of production is often brought about by standardization.)[23]

This transition to managerial control of the means and methods of production is not necessarily bad, especially when viewed from the point of view of product standardization; in fact, it is required to ensure standardization. Indeed, since standardization can only be brought about when the will to stabilize is balanced by sufficient experience to establish practical agreement, the need to rigidly enforce standards is a logical consequence of any standardization effort. Anarchy obviously contradicts the very essence of standardization. However, when the principles of standardization, which were developed for product or technical homogeneity, are applied to the management of quality assurance systems (as is the case with the ISO 9000 series, for example), one must consider whether or not these principles contradict, or at least challenge, some of the tenets of total quality management, flexible manufacturing, reengineering, and customized production described by many quality and manufacturing gurus. Unfortunately, the flexibility that was present in the early (1987) version of the ISO 9000 series of standards has slowly been eroded as more and more requirements (and thus increased pagination) are added under the pretense that such standards are demanded by phantom customers when in fact they are imposed by a small group of megalomaniacs.

Another side effect, brought about by the various attempts to regulate industries, is that regulations cannot be achieved without the creation of a bureaucracy that consists of administrators and an army of assessors/enforcers whose job it is to verify or interpret that the regulations and standards are "correctly" applied by industrialists. Once created, the survival instinct of any bureaucracy is to ensure its immortality. This immortality is achieved by collecting fees from the population of newly regulated users and by continually attempting to expand the bureaucracy's sphere of influence. Bureaucracies, and particularly regulatory bureaucracies, are not known for their efficiencies and swiftness of action. Among the myriad of examples provided by James Bovard in his *Lost Rights*, one of the

---

[22]Antoine Mal quoted in Jacques Ellul, *The Technological Society*, Vintage Books, New York, 1964, pp. 11–12.

[23]Ibid., p. 213.

most outrageous ones is the case of the U.S. Food and Drug Administration whose delays, Bovard contents, "are destroying the American medical device manufacturing industry." Medical devices that had already been approved for sale in Europe, Japan, and other countries were still awaiting FDA approval. A congressional report concluded that "[m]any of the small companies that populate the industry may be driven out of business altogether by regulatory delays. The process also means that Americans are denied health-care options that could be safer, more effective or less costly than those on the market today. . . ."[24]

In some cases, federal regulations can lead to some amusing headlines such as "Federal Laws Causing Toilet Shortage." The story, printed in a Seattle paper, explained that because of a new federal regulation aimed at conserving water, plumbers and contractors were having a difficult time finding the ultra-low-flow toilets that are to replace the 3.5-gallon tanks that were the standard for about a decade.[25]

Standards and the Law: A Powerful Combination

When, in their attempt to enforce standards unilaterally, nations transform voluntary standards into statutory requirements, the combination of standards and legal obligation is a potent one. In the United States, the association between standards and legal obligation has had a long history. In Europe, ever since the 1980s, the passage of hundreds of European directives, which are to be enacted by the various national legislations of member states of the European Community, duplicates the climate of legal enforcement of standards already found in the United States and exemplified by the activities of such government organizations as OSHA, the FDA, and the EPA. Is this necessarily good for society? If you answer "yes," I would suggest you read an excellent book by Philip K. Howard entitled *The Death of Common Sense*, and perhaps reconsider your answer.[26] Howard's main thesis is as follows: In their attempt to anticipate every possible circumstance, lawyers and government officials have over the years developed an incredibly complex and detailed legal system that consists of tens of thousands of rules and regulations which add up to hundreds of thousands of pages. In an attempt to manage this gargantuan system and to ensure fairness and due process of law, processes and procedures are constantly added to the already immense pool of procedures. This senseless unending expansion of procedures has led to what Howard calls the "Process Paradox," which simply states that, in order to act on anything, the government must first follow seemingly endless procedures. The cost of enforcing such a complex system is not only mind boggling, but more importantly, impossible to manage rationally and consistently, hence, the death of common sense where the ritualistic applications of rules and

---

[24]James Bovard, *Lost Rights*, St. Martin's Press, New York, 1994, p. 63. The cases cited by Bovard have been verified by the author while auditing medical device firms.

[25]*Daily Journal of Commerce* (Seattle, Washington), July 1994, p. 5.

[26]Philip K. Howard, *The Death of Common Sense*, Random House, New York, 1994.

regulations has become an unfortunate substitute for thinking. As Howard observes:

> *Look up at the Tower of Babel we are erecting in worship of perfectly certain and self-regulating authority. It admits no judgment or discretion; that, indeed, is the mortal sin it exists to eradicate. No one should ever, never ever, be allowed to exercise discretion: In matters of regulation, law itself will provide the answer. Sentence by sentence, it prescribes every eventuality that countless rule writers can imagine. But words, even millions of them, are finite. One slip-up, one unforeseen event, and all those logical words turn into dictates of illogic.*[27]

The thesis developed by Mr. Howard is very significant to the current international standardization movement. Meanwhile, I would suggest that all committee members who are busily writing more and more international guidelines and standards read Howard's book. Perhaps it will help bring some sanity to the current movement.

Standards Proliferation in an Age of Regulation

Encouraged by the unanticipated success of ISO 9000, the British Standards Institution could not resist the temptation of trying to launch a new registration program. The TickIT certification scheme was heavily promoted in 1993 to supposedly satisfy the need for quality assurance in the software industry. As had been the case with any marketing promotion before or since (e.g., Malcolm Baldrige), promoters always emphasized how the TickIT program was absolutely essential. After reading brochures praising the virtues of the TickIT certification program, one was left with a sense that software quality assurance had been recently invented by BSi. In reality, casual reference to historical events revealed that software quality assurance has been a well-established topic (at least in the United States) since the early 1960s. Fortunately, and with the possible exception of the United Kingdom where it first originated, one no longer hears much about TickIT. The software industry, for the most part, wisely rejected it.

In the last few years the International Organization for Standardization has managed to produce an unending list of guidelines and technical as well as non-technical standards. After the ISO 9000 series came the ISO 14000 series for environmental management, which already has a long list of guidelines. Efforts to produce yet other standards in health and safety management were voted down by U.S. representatives not so much because they did not like the idea but, more likely, because they did not want the new international standard to encroach on OSHA's domain. By the year 2000, the 9000 series of standards will have been reorganized and significantly rewritten (ISO 9001–2000) and, of course, expanded to include new requirements and a brand new structure. This new and improved ISO 9001 standard may keep a lot of training organizations busy for a few months

---

[27]Ibid., p. 51.

unless of course, the public finally catches on and refuses to go along for the ride (which is unlikely). As always, the market, as controlled by the biggest firms, will determine what is or is not needed. Still, a market (small as it may be) might develop for a long enough period of time for the ISO 9000 training industry to thrive once again for a while.

Of course, this proliferation of standards is not necessarily a unique phenomenon of the 1990s. More than 100 years ago, a nearly identical movement was under way. During the 1880s and prior to that, major fires and other calamities such as structural failures had caused several deaths in major U.S. cities. This prompted the emergence of a movement calling for the creation of building laws "devised for the public good." In 1891, the collapse of the Taylor Building in New York prompted Edward Henry to write an article titled "What Is the Use of Building Laws?" In it, Henry lamented on the limited value of the building inspector:

> *This dreadful calamity might have been prevented had the building been duly inspected; but inspection, we are told, is out of the question. Not only is the number of inspectors inadequate for the duties imposed upon them, not only are they themselves incapable of doing what they have undertaken, not only are they without the means of enforcing the decisions they form from this inspection, but the condition of modern trade, the rapidity with which buildings are filled and emptied, submitted to strain and released from its renders the work of the inspector valueless.*[28]

What is very different about the debates of 100 years ago regarding the need for building codes and the current push for ISO 9000, 14000, and other standards is that today, dissenting opinions regarding the issues of ISO 9000 registration are very difficult to find. Back in 1891, *Engineering Magazine* did publish counterarguments. John Beverly Robinson, the voice of opposition, asked some pertinent questions in his "Why I Oppose Building Laws. A Rejoinder." Robinson resented the loss of liberty and saw the government-imposed requirement to build fire escapes as nothing more than a stampede toward socialism. "Each year," he warns, "the promise of government extends." Who would pay for all these inspectors, Robinson asked? And finally, "Why should he (Henry) make people who don't want buildings inspected pay for his gratification?"[29] Because cost–benefit analyses were probably never conducted, one will never really know how effective the fire escape program was. How many lives were truly saved thanks to the availability of fire escapes (which are no longer obligatory)?

Today, such debates are generally not available. Major U.S. periodicals in the field of quality, including newsletters, have, with few exceptions, carefully avoided printing articles questioning the ISO 9000 stampede. The cult-like chorus of

---

[28]Edward Henry, "What Is the Use of Building Laws?," *The Engineering Magazine*, II(2), 1891, p. 240.

[29]John Beverly Robinson, *The Engineering Magazine*, II(2), 1891, pp. 250, 251. Naturally, one cannot help but wonder how many people were actually saved thanks to the new fire escapes. Such statistics would probably be very hard to uncover.

praise is unanimous, to the point of being boringly predictable. The fear of losing business or of not being competitive is a favored argument forwarded by many as a justification for wanting to achieve ISO 9000 registration. Christopher L. Bell and James L. Connaughton, for example, state within the opening paragraph of their "International Environmental Standards" article that "[f]ailure to conform to standards adopted in this area *could severely restrict trade* for companies that manufacture or sell products or services abroad."[30] Fortunately, the authors do not state that failure to comply *will severely restrict trade*, still the element of fear has been implanted and, as the late philosopher Hans Jonas postulated, fear can induce responsibility.[31] Moreover, the reader is never really told how trade restriction regarding a firm's management of its environmental policy may apply. Although pollution certainly knows no frontier, one wonders why (or how) a European, or any other government would want to impose trade restrictions on a U.S. firm simply because it failed to register its U.S. plants to the ISO 14001 standard.[32] Nevertheless, incredible as this may seem this is precisely the scenario that some individuals would like to see unfold in the early part of the 21st century. In fact, what happened during 1999 is that although some U.S. agencies such as the EPA are favorable toward ISO 14001 and have even helped fund some programs in a few states, the EPA is careful to state that it is not requiring companies to achieve ISO 14001 registration. The push, weak as it is, to achieve ISO 14001 registration seems to actually be coming from a few larger corporations that already have achieved or are in the process of achieving ISO 14000 certification. In a strategy that is reminiscent of the early 1990s when companies would require their suppliers to achieve ISO 9000 certification, several companies, such as IBM, for example, are now requiring or strongly suggesting that their suppliers also achieve ISO 14001 certification.

Will It Ever End?

What is rather remarkable about the proliferation of international standards is that international committees first published a standard that addressed quality issues of a product (ISO 9000, 1987), then came the environment (ISO 14000, 1996), and finally came the failed attempt at producing a health and safety standard. The progression is revealing: (quality of) product, (quality of the) environment, and lastly an attempt at addressing the quality of work life. The need to address the elemental quality of work life has not been totally ignored, however. In October 1997, the New York-based Council on Economic Priorities Accreditation Agency published its Social Accountability 8000 (SA 8000) standard, which attempts to address human rights at the corporation level.

---

[30]Christopher L. Bell and James L. Connaughton, "International Environmental Standards," *Journal of Environmental Law and Practice*, no date, p. 1. The threat is repeated in the conclusion: "... companies will ignore these standards at their peril," which may very well be true.

[31]Hans Jonas, *The Imperative of Responsibility*, 1979.

[32]And yet, the U.S government and even the state of California have never hesitated to try to enforce its laws on other nations, especially tax laws.

Based on the unending stream of technical documents, guidelines, and standards produced by numerous prolific committees, one can only conclude that the eruption of standards documents is not likely to end. What may likely end is the public's appetite for such standards. As of mid-1998, it would appear that the public's appetite for international standards had reached satiation.[33]

Conclusion

As we recognize the need and value for product and process standardization, along with all of its side effects, as dictated by the constraints of a mass production and mass consumption environment, should we necessarily assume that such principles can be applied to the maintenance and management of quality assurance systems for all industries? In other words, should the regulatory concepts of standardization, developed since the 1920s to apply the principles of product manufacturing/assembling and dimensions, be applied to the quality systems designed to manage the means of production?

Is it possible, beneficial, and economical for society as a whole to attempt to regulate and standardize not only the products manufactured by industries but also the quality assurance systems supposedly required to produce these products? Is it true, as is often proposed, that the formula for the economic success of a firm is solely, or even mostly, dependent on the proper management of quality? If so, is it logical to presume that the management of quality can be neatly packaged into a series of international standards or quality awards, a sort of metamodel as some have suggested?[34] More importantly, why should such standards even be produced in the first place? Who are the interested parties: companies, consumers, or the regulators? Finally, is the standardization of quality assurance, derived in the 1980s, relevant to the corporate needs of the next decade? The next chapter attempts to answer these questions.

---

[33]See Amy Zuckerman's "58 Multinationals Question ISO 9000 Registration; NIST Seeks Standards Summit," *Quality Progress*, August 1998, pp. 16–21.

[34]Mustafa V. Uzmuri, "Management System Metamodels: The Impending Standardization of Management Practice," unpublished paper, Department of Management, College of Business, Auburn University.

CHAPTER 12

The ISO 9000 Phenomenon and the Privatization of Military Standards

A lie can travel halfway around the world while the truth is putting on its shoes.

—Mark Twain

A committee is a cul-de-sac down which ideas are lured and then quietly strangled.

—Barnett Cocks

No grand idea was ever born in a conference, but a lot of foolish ideas have died there.

—F. Scott Fitzgerald

Was There Quality before ISO 9000?

As I write these lines, I am listening to the hi-fi sound of a German made Telefunken radio that my father purchased in the early sixties. Although some of the tubes were changed once around 1992 and the turntable was fixed for very minor repairs during the same time, the radio still delivers through its apparently ageless speakers the same high-quality sound it did nearly 40 years ago. The radio's design, which demonstrates an affinity for simplicity and practicality, is a classic example of European functional design that focuses on what students of industrial engineering have referred to as "designing from the inside out."[1]

---

[1]For an excellent description of the history of design see Christopher Lorenz's *The Design Dimension*, Basil Blackwell, New York, 1986, and John Heskett's *Industrial Design*, Oxford University Press, New York, 1980. As Lorenz (*The Design Dimension*, p. 11) explains, "From the very start, a yawning gulf developed between European and U.S. conceptions of industrial design: the one highly intellectual and dedicated to functional simplicity (what has often been described as 'working from the inside out'), the other a styling tool at the service of sales and advertising, where the exterior was all-important, and the inside mattered little."

Founded in 1903, the Telefunken company still exists today. During the 1900s, Telefunken engineers were responsible for many inventions: an around the world broadcast in 1918, early experiments with television in 1928, and stereophonic recording in 1953. To this day, and despite the dominance of Japanese firms in the small radio industry, the company is still recognized as a leader in electronics and electronic components.

The engineers that designed the radio purchased by my father in the early 1960s did so long before ISO 9000 came into existence, which in my opinion clearly demonstrates that superior product quality that was possible before the ISO 9000 standards were ever produced. Quality and reliability without mission statements or even a quality policy! This simple fact, which could be reaffirmed with countless other stories, seems to me undeniable and, yet, when I tell the story to quality professionals who seem enamored with the ISO 9000 standards, I am told that before 1987 (the date the ISO 9000 series of standards first appeared), the quality of products left a lot to be desired. One individual, for example, reminded me that during the early 1980s, his parents had bought a defective car (colloquially known as a lemon). The argument was supposed to convince me that thanks to ISO 9000 no more lemons are produced! Of course, that argument is totally irrational. Although it would seem that fewer "lemons" are produced than 10 years ago, the improvement in quality has nothing to do with the emergence of the ISO 9000 or the automotive industry's even more demanding QS 9000 standards. American cars have improved because of economic necessity; it was a simple case of do or die. Moreover, the improvements were introduced long before ISO 9000 (or QS 9000) was first published in 1987. Had the industry waited until 1987 it may not have survived the crisis.

One cannot deny that the arrival of the ISO 9000 standards has done little to improve the overall quality of products, yet many quality professionals will certainly deny that fact with unsubstantiated statements of praise for the ISO 9000 series of standards. Of course, numerous articles have been written to unscientifically "prove" that ISO 9000 certification has been equivalent to economic salvation or a near religious experience. Enormous amounts of progress and savings are claimed by authors who swear that ISO 9000 saved their organization. The phenomenon is not new. Well over 30 years ago when yet another quality panacea (known as zero defects or ZD) was consuming quality professionals and countless industrialists, the wise Dr. Juran lamented that since companies tend to be carried away "by their own propaganda," no one will ever admit defeat.[2]

Are concrete facts, statistics from before and after ISO 9000 certification, for example, ever produced? I have never seen any. Moreover, can one truly expect anyone who has spent several months (and much money) implementing an ISO 9000 quality assurance system to admit that the system has not lived up to its expectations? Of course not. Also, would one expect magazines and periodicals that earn a significant portion of their revenues from ISO 9000 advertisers to publish any opposing views regarding the dubious benefits of ISO 9000 certifi-

---

[2]J. M. Juran, "Quality Problems, Remedies and Nostrums," *Industrial Quality Control*, June 1966, p. 653.

cation? Of course not, which is why few articles are published questioning the virtues of ISO 9000. Instead, what happens is that other fads and quick fixes are introduced (e.g., six sigma, SPC, Malcom Baldrige, etc.). Articles where the author (usually a quality manager) uses hyperboles to convince the reader that ISO 9000 saved his company have always surprised me. If I were the owner of the company, I would like to know how much improvement was seen as a result of the costs expended.

Still, the ISO 9000 series of standards has benefited many companies. The type of company that is most likely to benefit from the rigor of ISO 9000 certification has been described by Tom Juravich in his *Chaos on the Shop Floor*. Describing the working condition at National (a fictitious name), Juravich writes: "What I saw at National, however, was a great deal of confusion and chaos even in very simple jobs; procedures were changed, machines broke down, poor materials were pushed through the line. Much of this confusion resulted from decisions made by management. Most decisions were made for the short run, without long-term goals in mind. But more than that, some decisions seemed to have no goal at all and could only be perceived as 'bad' decisions."[3] For companies like National where training employees is never considered a wise financial option, where equipment maintenance is anathema, and where standardized procedures are inconceivable, the implementation of an ISO 9000 quality management system would certainly be a great benefit *if and only if* one could somehow convince management to invest in such a venture.

Many companies have benefited from an intelligent and practical application of ISO 9000 but, unfortunately, based on my sample of ISO 9000-certified companies such success stories seem to apply to a minority rather than a majority of companies and I would suggest that the percentage of successful and beneficial ISO 9000 applications does not exceed 35 to 40 percent. Where did this ISO 9000 phenomenon come from and why have so many spent so much on its implementation?

Antecedents to the ISO 9000 Movement

During the past 50 years, American businessmen have seen numerous managerial and quality improvement fads come and go.[4] The 1930s and 1940s saw the emergence of receiving inspection, acceptance sampling, and control charts. During the 1950s, a period still very much influenced by Taylor's efficiency studies, Drucker's management by objectives, and the wonders of general systems theory, it was generally believed that most everything could be solved, improved, or fixed if only the proper technique was applied. In the field of quality control,

---

[3]Tom Juravich, *Chaos on the Shop Floor. A Worker's View of Quality, Productivity, and Management*, Temple University Press, Philadelphia, 1985, p. 14.

[4]The Frenchman Charles Bedaux who created the B unit in the 1930s became very popular in the United States. By 1942, 720 corporations with 675,000 workers had adopted the Bedaux system. Today, the Bedaux system has long been forgotten and replaced by countless other fads. See Bedaux's Labor Management, N.Y. 1928, cited in Siegfried Gideon, *Mechanization Takes Command*, Oxford University Press, New York, 1948, pp. 114–115.

for example, statistical process control techniques developed in the 1920s continued to be very popular during the 1950s and 1960s, and even to this day.

The proliferation and reliance on techniques used by organizations to solve their daily problems create what sociologists have referred to as a dualistic society. Quoting the French sociologist Antoine Mas, Jacques Ellul observed that "the intervention of a technique of organization results in the creation 'of two classes very far removed from one another. The first, numerically small, the second, infinitely more numerous, is composed of mere executants. . . .' The latter are hacks who understand nothing of the complicated techniques they are carrying out."[5] Practitioners of these techniques, Ellul tells us, create a kind of closed fraternity. Ellul's observations are certainly valid and pertinent when one considers the variety of statistical techniques (statistical process control and design of experiment or DoE, for example) applied and misapplied daily by thousands of amateur statisticians.[6]

Similarly, the use of psychological tests so popular in the 1950s and still used to this day by some major consulting firms helped promote the concept of specialists. Reviewing William H. Whyte's description of the "psychological phenomenon," one cannot help but notice the striking similarities with the ISO 9000 phenomenon of the 1990s. Describing how, during the 1950s, Sears, General Electric, Westinghouse, and others made use of psychological tests, Whyte observes that "[G]iving them has become something of an industry itself. In the last five years the number of blank test forms sold has risen 300 percent. The growth of psychological consulting firms had paralleled the rise."[7] This unprecedented growth, which closely resembles the pattern of ISO 9000 registration experienced during the early 1990s (see Figure 12–1), led to a commensurate growth in consultation agencies and consultants. It was not long before a program designed to *regulate* consultants was approved. Describing a process that reminds us of the current ISO 9000 phenomenon, Whyte noted that some agencies reported that in one year, "seven hundred new consultants asked to be put on the *approved list of customers*. Colleges are also getting into the business . . . a king of competition, incidentally, which annoys a good many of the frankly commercial firms."[8] (*Note:* Colleges offer ISO 9000 seminars and, during a brief period, certain factions asked that the American Society for Quality or the Registrar Accreditation Board should control the growth of ISO 9000 consultants by requiring them to register their services as is currently done in Malaysia, for example.)[9]

During the 1960s, some quality professionals began to promote the zero-defect movement. Few professionals recognized the limitations of the ZD movement

---

[5]Jacques Ellul, *The Technological Society*, Vintage Books, New York, 1964, pp. 162, 275.

[6]As early as 1953, Professor Grant believed that "the use of statistical techniques should be reserved for a small class of professional specialists." Eugene L. Grant, "A Look Back and a Look Ahead," *Industrial Quality Control*, July 1953, pp. 31–35, quotation is from p. 35.

[7]William H. Whyte, Jr., *The Organization Man*, Doubleday Anchor Books, New York, 1956, p. 193.

[8]Ibid., p. 193.

[9]*ISO 9000 News*, 3(3), May/June 1994, pp. 18–20.

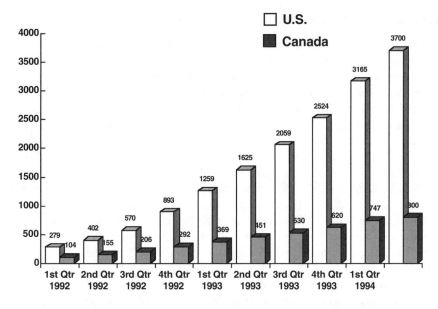

FIGURE 12–1 ISO 9000 Registration by Quarter (1992–1994)

and even fewer wrote about them, although J. M. Juran warned that ZD programs had been conducted mainly "by companies which are under contract to the government military and space agencies." Juran also questioned the validity of the many claims of improvement achieved supposedly thanks to the implementation of a ZD program. In the rare instances where qualitative data were published, Juran concluded that the "published qualitative statements of results were highly colored versions of the actual results." Juran astutely observed that since people simply do not report their failures, only success stories get to be published. Recognizing that ZD-type programs could be of value in certain "narrow applications" where operator controllability has a good deal of validity (such as wiring or soldering, for example), Juran nonetheless pointed out that "[I]t is tempting to dismiss the ZD type program as a new form of hokum, cooked up by publicity seeking quacks to build up their public image, to sell their services and books, to become the high priests for a new religion, to become kings of a newly created hill."[10] The 1970s and 1980s had their total quality movements, many imported from Japan (quality circles, more SPC, design of experiment, Just-in-Time manufacturing, etc.), including the quality award phase (Baldrige, Quality Cup, state quality awards). Yet, despite Juran's warning, the preaching has not stopped and continues with the ISO 9000 movement. Why?

---

[10]Juran's quotations are found in his "Quality Problems, Remedies and Nostrums," *Industrial Quality Control*, June 1966, pp. 647–653.

The Need for Quick Fixes

As we review the long list of movements that have come and gone (or lingered on) during the past decades, one wonders why they are so popular and how they are successfully and continuously being replaced or recycled year in and year out. There are certainly many factors. Juran's reference to the "high priests for a new religion," true as it was in the mid-1960s, is certainly even more apropos during the 1990s. A brief review of the field of quality consultation during the past 50 years reveals a steady increase in the number of quality consultants. In the 1940s, no more than half a dozen consultants advertised their services in *Industrial Quality Control*. In 1968, the year *Quality Progress* published its first issue, the number of quality consultants advertising their services had risen to 16 to 18. By 1994, the number of advertisements had risen to 70, most of them claiming some expertise in ISO 9000. Obviously, the number of consultants nationwide far exceeds this number and, as far as ISO 9000 consultants are concerned, the number has been estimated to be anywhere between 1200 and 1500.[11]

But these providers of "quality services" could not survive if a market for quick and easy fixes did not exist. Michael McGill, who has analyzed the phenomenon from the management fads point of view, explains that "the very processes of science—the deliberate (slow) search for specific (not universal) alternatives (not solutions) to be administered by experts (not managers) in complex (not simple) situations—flies in the face of what managers want."[12] This phenomenon, McGill observes, has led to the growth of a "quick fix" industry that promotes rapid and simplistic technical solutions summarized in carefully produced glossy brochures.[13] Shopping for the perfect course, brochure hunters are then faced with the difficult task of selecting the "brochure philosophy" that they think best suits the needs of their company. The selection is almost endless: employee empowerment training, statistical training, just-in-time training, quality circles, MBO, T-group (awareness groups), operation research, quantitative management, PERT/CPM, process simulation techniques, gauge R&R, Malcolm Baldrige preparation training, and, more recently, ISO 9000 lead assessor training, ISO 9000 registration, as well as the plethora of ISO 9000-related training and seminars that has led to the "ISO 9000 phenomenon." The chosen techniques are often offered in a modular fashion and usually with little if any consideration as to their appropriateness to the industry.[14]

---

[11]See "The Craze for Consultants," *Business Week*, July 25, 1994, pp. 42–45.

[12]Michael McGill, *American Business and the Quick Fix*, McGraw-Hill, New York, 1988, pp. 31–32. "[T]here is truly a 'quick fix' industry and each fix is a product with attendant market strategy, add-ons and spin-offs, and franchises. There seems to be an attitude among managers that if the message isn't packaged for mass marketing, it can't be worth very much" (p. 28).

[13]Ibid. "Fads and fixes emerge from the myths of management and, in turn, promote the same myths. The result is a managerial morass wherein simplistic solutions take form, flower briefly, then sink back to feed new forms. So long as management stays mired in its own mythic-ridden marshland, effective substantive solutions will evade managers, trapped as they are by their own imagined constraints and by objectives chosen by habits" (pp. 5–6).

[14]With respect to the inappropriateness of certain techniques, McGill relates how paper weight followed as the military began to adopt and promote these techniques (PERT/CPM). Wright-Patterson Air Force Base, for example, had PERT charts everywhere and despite the enormous overhead cost

As training agencies begin to offer public or in-house seminars, a series of facts and occasional half truths are slowly but effectively dispensed and dutifully absorbed by hundreds and, eventually, thousands of trained individuals. Within a short time, a subculture of self-proclaimed experts is born; a phenomenon is about to happen.

The ISO 9000 Phenomenon: A Case Study in the Manufacturing of Consent

The expression "manufacturing of consent" is taken from the title of a book by Edward S. Herman and Noam Chomsky. In it, the authors suggest that "the 'societal purpose' of the media is to inculcate and defend the economic, social, and political agenda of privileged groups that dominate the domestic society and the state. The media serve this purpose in many ways: through selection of topics, distribution of concerns, framing of issues, filtering of information, emphasis and tone, and by keeping debate within the bounds of acceptable premises."[15] I propose that, with regard to the ISO 9000 phenomenon, a series of events and principles similar to those explored by Herman and Chomsky have been at play.

The ISO 9000 Series

The ISO 9000-9004 series officially came into existence in 1987 when the following five documents were published by the International Organization for Standardization in Geneva[16]:

1. ISO 9000-1 (ANSI/ASQC Q90), *Quality Management and Quality Assurance Standards: Guidelines for Selection and Use*
2. ISO 9001 (ANSI/ASQC Q91), *Quality Systems—Model for Quality Assurance in Design/Development, Production, Installation, and Servicing*
3. ISO 9002 (ANSI/ASQC Q92), *Quality Systems—Model for Quality Assurance in Production and Installation*
4. ISO 9003 (ANSI/ASQC Q93), *Quality Systems—Model for Quality Assurance in Final Inspection and Test*
5. ISO 9004-1 (ANSI/ASQC Q94), *Quality Management and Quality System Elements Guidelines.*

Ford implemented the system. These "formulas" failed, McGill states, because they were too simple. McGill, *American Business*, pp. 11–12.

[15]Edward S. Herman and Noam Chomsky, *Manufacturing Consent: The Political Economy of the Mass Media*, Pantheon Books, New York, 1988, p. 298. The author admits to borrowing the expression from the journalist Walter Lippman who used it in the 1920s.

[16]The International Organization for Standardization traces its origins to the International Federation of the National Standardization Association (1926–1939). From 1943 to 1946, the United Nations Standards Coordinating Committee (UNSCC) acted as an interim organization. In October 1946 in London, the name International Organization for Standardization was finally agreed on. The organization, known as ISO, held its first meeting in June 1947 in Zurich. See Howard Coonley, "The International Standards Movement," in Dickson Reck (Ed.), *National Standards in a Modern Economy*, Harper & Brothers, New York, 1954, Chap. 5.

Although not immediately accepted by the world community, since 1989 the standards' popularity has increased exponentially.[17] In the United States, where the standards have been adopted as the ANSI/ASQC Q-9000-9004 series, nearly 3000 firms had achieved registration as of December 1997. Perhaps the single most important driving force that has led to today's guarded acceptance of the ISO 9000 series has been the fear of being locked out of the European market once economic unification is completed. This fear, exaggerated as it may be, can, in part, be attributed to occasional inaccurate press coverage.

Origins of the ISO 9000 Standards

Tracing the origins of standards is not an easy or rewarding task. Recounting their origins is even more difficult because one must first describe the standards and then demonstrate the similarities with previous standards (see the discussion of military regulations in Chapter 9). The similarities between military standards and the ISO 9000 series are rather striking (see Table 12–1). What is remarkable is that the resemblance between the two documents transcends the similarity in phrasing. Thus, when one compares the practices of ISO 9000 implementation and registration with the quality systems that were developed by government contractors to address MIL-Q-9858A and contrasts the auditing practices of the government, which were designed to ensure that these "quality programs" were in fact effective, one notices that everything is being reinvented and applied to the private sector.

MIL-Q-5923 and 9858 and ISO 9000: Déjà Vu!

I realize that many individuals who have either recently discovered the ISO 9000 series or were exposed to it in the late 1990s probably think that the ISO 9000 series is the greatest thing since sliced bread. Still, many veterans, particularly those of you who have had to deal with MIL-Q-9858A (or the Air Force's 5923 published in December 1950) during the past 35 to 45 years, have long known that there is really nothing new about the 9000 series: It is certainly an old concept that has been ever so slightly updated and essentially repackaged for consumption in the 1990s. However, being aware of the fact that there might be a few skeptics, I will briefly review an article that appeared more than 35 years ago, in the January 1963 issue of *Industrial Quality Control*.

In an article titled "Does Your QC Organization Meet MIL-Q-9858?," Harold L. Kall of the Carborundum Company provided a detailed review of the MIL-Q-9858 evaluation process. Kall explains:

> *Upon acceptance by our company of a contract where MIL-Q-9858 was a mandatory specification, a request was made for an evaluation of the quality control system for compliance with the specification, to the government inspec-*

---

[17]By the end of 2000, the three "auditable" standards, ISO 9001, 9002, and 9003, will be reorganized and reduced to only one standard ISO 9001-2000.

TABLE 12–1 Partial Comparison of MIL-Q-9858A Text with ISO 9001-1994 Text

MIL-Q-9858 (citations are partial)	ISO 9001-1987 (citations are partial)
3.4 Records The contractors shall maintain and use records or data essential to the economical and effective operation of his quality program. These records shall be available for review by the Government Representative and copies of individual records shall be furnished him upon request. Records are considered one of the principal forms of objective evidence of quality. The quality program shall assure that records are complete and reliable.	4.16 Quality Records The supplier shall establish and maintain procedures for identification, collection, indexing, filing, storage, maintenance, and disposition of quality records. Quality records shall be maintained to demonstrate achievement of the required quality and the effective operation of the quality system. . . . Where agreed contractually, quality records shall be made available for evaluation by the purchaser or the purchaser's representative for an agreed period.
Inspection and testing records shall, at a minimum, indicate the nature of the observations together with the number of observations made and the number and type of deficiencies found.	4.12 Inspection and Test Status The inspection and test status of product shall be identified . . . [to] indicate the conformance or nonconformance of product with regard to inspection and tests performed. (*Note:* The reference to records is mentioned in at least a dozen paragraph including receiving, in-process, and final inspection.)
Also, records for monitoring work performance and for inspection and testing shall indicate the acceptability of work or products and the action taken in connection with deficiencies.	4.13.1 Nonconformity Review and Disposition The description of nonconformity that has been accepted, and of repairs, shall be recorded to denote the actual condition.
The quality program shall provide for the analysis and use of records as a basis for management action.	4.1.3 Management Review The quality system adopted to satisfy the requirement of this standard shall be reviewed at appropriate intervals by the supplier's management to ensure its continuing suitability and effectiveness. Records of such reviews shall be maintained.
6.4 Handling, Storage and Delivery	4.15 Handling, Storage, Packaging, and Delivery (Very similar to the phrasing found in 9858A)
6.5 Nonconforming Material The contractor shall establish and maintain an effective and positive system for controlling nonconforming material, including procedures for its identification, segregation and disposition. All nonconforming supplies shall	4.13 Control of Nonconforming Product The supplier shall establish and maintain procedures to ensure that product that does not conform to specified requirements is prevented from inadvertent use or installation. Control shall provide for identification,

TABLE 12–1 *(Continued)*

MIL-Q-9858 (citations are partial)	ISO 9001-1987 (citations are partial)
be positively identified to prevent unauthorized use, shipment and intermingling with conforming supplies.	documentation, evaluation, segregation when practical, disposition of nonconforming product, and for notification to the functions concerned.
5.2 Purchasing Data The contractor's quality program shall not be acceptable to the Government unless the contractor requires of his subcontractors a quality effort achieving control of the quality of the services and supplies which they provide.	4.6.2 Assessment of Sub-contractors The supplier shall select sub-contractors on the basis of their ability to meet sub-contract requirements, including quality requirements. The supplier shall establish and maintain records of acceptable sub-contractors. Other paragraphs (4.6.3, for example) follow very closely the requirement for data verification found in 9858.
6.3 Completed Item Inspection and Testing 4.2 Measuring and Testing Equipment	4.10.3 Final Inspection and Testing 4.11 Inspection, Measuring and Test Equipment
6.6 Statistical Quality Control and Analysis	4.20 Statistical Techniques Specific reference to various sampling plans found in 9858A are not included in paragraph 4.20.

tion agency that had been delegated this responsibility by the contracting officer. Copies of the Quality Control Manual were supplied to the inspection agency prior to the inspection, and a mutually acceptable date for the evaluation was then established.[18]

Individuals who have been through an ISO 9000 audit or assessment will recognize the striking similarity in process. I have emphasized the word "evaluation" because I believe the word should be adopted by ISO 9000 registrars. Rather than refer to ISO 9000 audits, one should use the word "assessment" or "evaluation," which better reflects the purpose of the registration exercise. Indeed, as Kall eloquently explained more than 35 years ago, the 9858 evaluation process should perhaps be rediscovered by some contemporary auditors:

The government evaluation team conducted a short review with members of top management to acquaint them with the purposes and objectives of MIL-Q-9858 and to describe the exact procedure they planned to use to evaluate the quality control system. Following this interview, individual members of the team inspected specific areas of our system. The emphasis was not on how

---

[18]Harold L. Kall, "Does Your QC Organization Meet MIL-Q-9858?," *Industrial Quality Control,* January 1963, p. 11.

something should be done, but rather on how effectively the system operated and whether it was adequate to comply with the contract. Particular
emphasis was placed on whether the written program was actually being followed in the various areas of the plants.[19] *[Emphasis added]*

Kall goes on to briefly explain other key activities of the government team,
namely: plant tours, exit interviews, and follow-up (identical to today's
approach!).

After presenting the table of contents of a quality control manual suitable for
MIL-Q-9858 (which would almost be suitable for today's ISO 9002-1994), Kall
concludes with some valuable comments that are still very pertinent today:

*The government does not desire to suggest that any particular organizational
set-up or quality control procedure must or should be used in your plant. This
decision is an individual company responsibility. In view of the increasing
complexity of the products being purchased by the government, the quality
assurance programs must be adequate to meet very tight quality requirements.
However, the existence of a good quality control program on paper is not sufficient: the program must be used.[20]*

Not surprisingly, the ISO 9000 series of standards has adopted a similar philosophy that, unfortunately, is not always practiced by all registrars. As the reader will
observe, the following paragraph, taken from the introduction of the ISO 9001
standard, closely parallels Kall's observations:

*. . . it is not the purpose of these . . . standards to enforce uniformity of quality
systems. They are generic and independent of any specific industry or economic sector. The design and implementation of a quality system will be influenced by the varying needs of an organization, its particular objectives, the
products and services applied, and the processes and specific practices
employed. [These standards] may need to be tailored by adding or deleting
certain quality-system requirements for specific contractual situations.[21]*

There certainly is much truth to the proverb "The more things change, the more
they remain the same."

Evolution of the ISO 9000 Movement in the United States

The earliest U.S. reference I could find that specifically alludes to the need for
an international system of quality assurance is in an abstract written in 1976 by
J. G. Gaddes, then director of the British Standards Institution, and entitled

---

[19]Ibid., p. 11.

[20]Ibid., p. 12. Remember that Griffith's account already mentioned that 10 years earlier the same
principles were practiced by the Air Force inspectors.

[21]ANSI/ASQC Q9001-1994, p. vii. ANSI/ASQC Q9001 is the American equivalent of the ISO 9001
standard.

"International Certification from the U.K. Point of View."[22] In the abstract Gaddes proposes that since the United Kingdom is dependent on international trade (which industrial country is not?), an international certification system would be of particular significance. (However, we are not told why such an international scheme would be beneficial and to whom.) Gaddes concludes his abstract by stating that "[T]here will be particular references to the relevant work of the International Organization for Standardization (ISO). . . ." In the same *Transactions*, Mr. P. Corner, then of the U.K. Ministry of Defense, proposes a similar argument and writes of the need to have international standards "which produce first class quality and an awareness of the need to apply these to all their endeavors."[23] How the use of international standards would guarantee first-class quality is never really explained by Mr. Corner. Yet the following quotation found in an April 1988 issue of *BSi News* might explain why the British were so interested in actively promoting their BS 5750 parts 1–3 standards, which eventually were accepted (with some modifications) by Geneva's International Organization for Standardization as the ISO 9001, 9002, and 9003 standards:

> *Maximizing the benefits to British industry of the Single European Market of 1992 is now an* overriding concern of BSi *and of business and Government in the UK. Achieving acceptable standards is essential* if we are to win trading advantage from this market of 320 million customers. *Never has it been more important for British interests* to seize the opportunities and counter the threats inherent in a trade barrier free Europe. . . . Taking on the chairmanship or secretariat of key committees is expensive and time-consuming, but essential if the UK is to control the pace of work in areas of special importance to British industry. It is vital for the future trading success of the UK that we influence the decisions being taken in these areas.[24] *[Emphasis added]*

Another 5 years elapsed before Robert W. Peach, who had been promoting the ANSI/ASQC Z1.15 standard, *Generic Guidelines for Quality Systems*, reintroduced his audience to the idea of ISO and international standards.[25] The follow-

---

[22]J. G. Gaddes, "International Certification from the U.K. Point of View," in *ASQC 1976 Technical Conference Transactions*, ASQC, Toronto, Canada, p. 441. The paper was not submitted.

[23]P. Corner (U.K. Ministry of Defense), "Quality—A Common International Goal—The UK Approach," in *ASQC 1976 Technical Conference Transactions*, ASQC, Toronto, Canada, p. 436.

[24]Quoted from "Reaping the Benefits of the Single Market," *BSI News*, April 1988, p. 5. The BSI 5750 series, which borrows extensively, yet indirectly (via NATO's AQAP requirements), from various U.S. military standards, was eventually used as the model for the ISO 9000 series adopted in 1987 by the International Organization for Standardization. Certainly one cannot fault the British for wanting to try to control committees, particularly technical committees which, I am told, are often dominated by German delegations that are always very well prepared and usually outnumber their counterparts (one would hope that ANSI is equally aggressive). However, what is the significance of such a policy for U.S. manufacturers? More specifically, within the arena of ISO 9000 registration and seminars, one would like to know if there is an implicit or explicit link between BSi's stated policy (as exemplified by the above quotation) and the very successful program of British and worldwide certification/registration.

[25]Robert W. Peach, "International Quality System Standards," in *ASQC Quality Congress Transactions*, ASQC, San Francisco, CA, 1981, pp. 327–331.

TABLE 12–2 Members of the U.S. ISO/TC 176 Group

Richard A. Freund (Eastman Kodak Co.)
Harvey S. Berman (Underwriters Laboratories)
A. Blanton Godfrey (Bell Telephone Laboratories)
John L. Kidwell (?)
J. J. Lutzel (Allen-Bradley Co.)
Donald W. Marquardt (Du Pont Company)
August B. Mundel (Consultant?)
Robert W. Peach (W. A. Golomski & Associates)
Aaron B. Rosenthal (General Motors Corp.)
Lawrence A. Wilson (Lockheed-Georgia Co.)

ing year, Peach and the late Donald W. Marquardt (then of E. I. du Pont de Nemours & Co.) published two articles devoted to international standards and the workings of the Standards Technical Committee on Quality Assurance (TC 176), the committee in charge of writing what was to become the ISO 9000 series. Marquardt's article is interesting because Marquardt recognizes that a distinction had to be made between multilevel quality assurance standards that evolved from military procurement standards and the so-called "generic" quality standards that evolved from "marketing-oriented industries, where contractual quality assurance relationships between buyer and seller are absent or peripheral."[26] Recognizing the important differences between the two applications, Marquardt concluded by stating:

It is proposed that international standards being developed provide a guide for standards users to identify the environment in which they are operating in a given situation, and thus to select appropriately a marketing environment quality management system or a contractual/regulatory environment quality assurance system, or a hybrid of the two. The standards should be based on the distinction developed within this paper.[27]

Unfortunately, Marquardt's recommendations were either ignored or so diluted as to have lost any significance (although his recommendations appear to have been partly rediscovered in 1994). Indeed, the ISO 9000 series of standards does not distinguish between customers coming from a regulated industry or a government agency and the millions of customers who rarely if ever participate in contract negotiations with their "suppliers."

That this should occur was probably unavoidable when one considers the make-up of the first U.S. Technical Advisory Group on ISO 9000 formed in the very early 1980s (see Table 12–2). It is unlikely that representatives from Eastman

---

[26]Donald W. Marquardt, "Comparison of Multi-Level and Generic Standards" *ASQC Quality Congress Transactions*, 1982, ASQC, Milwaukee, pp. 244–248, and Robert W. Peach, "Applying an International Generic System Standard," *ASQC Quality Congress Transactions*, 1982, ASQC, Milwaukee, pp. 249–253.

[27]Ibid., p. 247.

Kodak, General Motors, Lockheed, Bell, Allen-Bradley, and so on, would have understood the need of the small businessperson. Most of these organizations had already implemented or were about to implement a supplier evaluation program. Government contractors were familiar with the military specifications, and the automotive industry (Ford, for example) had recently adopted a supplier surveillance program that was designed along the lines of military specifications. Supplier evaluation was a world very familiar to these committee members. Indeed they spoke the same language of quality.

The Universal Language of Quality

While it is true that committee members do not always agree with everything every other member has to say, they nonetheless share a common objective and are united by a common purpose: to generate standards and promote their diffusion. Moreover, since they tend to belong to the same professional group collectively known as quality professionals (or quality engineers) they naturally share a common culture of quality (one might even speak of an ideology of quality) and a common quality-speak. By quality-speak, I mean a specialized, and at times technical, language characterized by a unique jargon. Like all other professionals, quality-speakers have a propensity to use quality-speak acronyms or short expressions to express ideas, and the list is long: just-in-time, proactive, SPC, DoE, TQM, reengineer, change agent, six sigma, benchmark, FMEAs, mistake-proofing, fish bone diagram, root-cause, KPIs (key performance indicators), lead assessor, CARs, NCRs, CPKs, ISO this and ISO that, empowerment, and so on.

This ability to speak in quality-speak serves two purposes: On the one hand, it allows quality professionals to differentiate themselves from any other group and to "see" the world differently from others. As Benjamin Whorf observed long ago:

> [W]e dissect nature along lines laid down by our native languages. The categories and types that we isolate from the world of phenomena we do not find there because they stare every observer in the face; on the contrary, the world is presented in a kaleidoscopic flux of impressions that has to be organized by our minds—and this means largely by the linguistic systems in our minds. ... We are thus introduced to a new principle of relativity, which holds that all observers are not led by the same physical evidence to the same picture of the universe, unless their linguistic backgrounds are similar, or can in some way be calibrated.[28]

By using quality-speak, quality professionals perceive the world in their own unique way. Unfortunately, this ability to see the world through quality-speak

---

[28]B. L. Whorf, *Language, Thought and Reality*, MIT Press, Cambridge, MA, 1973, pp. 233–234. For a description of the Whorfian theory applied within the context of Mexican culture, see Chap. 4, "The Sons of the Malinche," in Octavio Paz's *The Labyrinth of Solitude (and Other Essays)*, Grove Press, New York, 1985. See also Alan Riding's Chap. 1, "The Mexicans," in his *Our Distant Neighbors*, Alfred A. Knopf, New York, 1985.

lenses has some serious side effects. Quality professionals who can only think in terms of quality-speak are seriously limited in their ability to offer creative thoughts. Indeed, quality-speak has many of the characteristics of George Orwell's Newspeak, which he describes in his famous novel, *1984*. "Newspeak," Orwell tells us, "was the official language of Oceania and had been devised to meet the ideological needs of Ingsoc, or English Socialism. . . . The purpose of Newspeak was not only to provide a medium of expression for the world view and mental habits proper to the devotees of Ingsoc, but to make all other modes of thought impossible . . . in Newspeak the expression of unorthodox opinions, above a very low level, was well-nigh impossible."[29] Replace "Oceania" with "the quality profession," "Ingsoc" with "quality," take out "English Socialism" and Orwell's sentence would apply just as well today to the quality profession, which is well known for its "Qualityspeak."

This ability to share a common language of quality is but one factor that can help forge what might be referred to as an ideology. Another important factor that must also be considered is that decisions and/or resolutions regarding the fate and content of any standard is essentially controlled by one or more national or international committees. Committees, as some researchers have observed, can have their drawbacks. Commenting on the rationality of committees and groups in general, Stuart Sutherland writes that studies have shown that "in addition to taking more extreme decisions than individuals, the members of a group are more confident about the correctness of the group's decisions than about that of their own."[30] This self-serving sense of solidarity is potentially dangerous because, as Sutherland observes, such extreme (i.e., riskier) committee decisions are often *less rational* than decisions reached by an individual.

Of course, the ISO 9000 series has to be approved by an international committee of experts who speak several languages and therefore one could argue that the Whorfian hypothesis does not apply in this case, but it does. The calibration referred to by Whorf is, to some extent, provided for by several factors. For one thing, most committee members review the drafts and offer comments in English. Second, many members have known each other for several years. Third, the draft documents are usually written by a handful of dedicated and committed "volunteers" who may originate from only two to three countries. The fourth and perhaps most important reason is that many of the influential members (that is, those that can afford to spend the time demanded by the various committees) are employed by major corporations, registrars, or official governmental organizations and, thus, tend to represent the interests of these corporations or organizations. Finally, most committee members seem to have at least one more common characteristic: They are, in all likelihood, used to dealing with bureaucracies and see nothing wrong with reproducing and perpetuating them, supposedly for the good of society. This is made clearly evident when one considers that the number of ISO 9000-related standards not only more than

---

[29]George Orwell, *The Orwell Reader*, Harcourt Brace & Company, New York, 1984, pp. 409, 410, 417.

[30]Stuart Sutherland, *Irrationality*, Rutgers University Press, New Brunswick, NJ, 1994, pp. 65, 66.

doubled during the late 1990s, but the trend is to also increase the pagination (the 45 pages for the 1987 ISO 9000-9004 series increased to 75 pages for the 1994 edition and will likely increase further for the 2000 revision). To make matters worse, the community of ISO 9000 promoters who proceeded to transform the ISO 9000 series into a fad and to advertise it reinforced the image of the ISO 9000 series as a compulsory requirement.

What Others Have Said about the Series

Some classic quotations about the ISO 9000 series or the registration process are included in Table 12–3 with the author's associated comments.

Evolution of the ISO 9000 Series

The interest in ISO 9000 registration, as demonstrated by the hundreds of articles written about the series during the 1992–1996 period, may have led anyone to believe that nearly every company in North America was about to be registered to one of the ISO 9000 standards. Yet the number of U.S. manufacturing firms that have achieved some form of ISO 9000 registration as of 1999 totaled but a fraction of a percentage point (certainly less than one-half of a percent) of the more than 15 million businesses operating in the United States. When one reviews the evolution of the current interest in the ISO 9000 series, one is struck by the fad-like nature of the phenomenon (see Tables 12–4 and 12–5).

A review of non-U.S. publications (Brazil, France, United Kingdom, Australia, Peru) reveals that similar marketing patterns of ISO 9000 propaganda developed overseas. (This is logical since most promotions are run by international registrars headquartered in Europe or the United States.) The message is clear: Achieve registration or you may lose your competitive edge. The continuous hammering of the ISO 9000 propaganda machinery is relentless—Table 12–5 breaks down the advertising trends and the number of ads published in *Quality Progress* alone. The number of advertisements and revenues solely generated by the advertising and/or promotion of the ISO 9000 phenomenon are staggering.

But, should we necessarily conclude that because a message is repeated several times by various organizations it is necessarily true? Of course not. In fact, in the United States as early as 1993, a few articles, including some written by the author, began to question the rationality of the ISO 9000 phenomenon and its emergent bureaucracy.[31] By mid-1996 it became clear that interest in ISO 9000 had trickled to almost nothing. Companies that once reached a rapid decision to start implementation procedures would drag the decision process along for several months only to conclude that "they are not now ready for implementation."

---

[31]A similar phenomenon occurred in the United Kingdom when in 1988, Allen J. Sayle created quite a stir by publishing his "ISO 9000—Progression or Regression?" *QA News*, 14(2), February 1988, pp. 50–53. Over the next few months at least 15 or more letters were published by the editor. The letters covered a whole range of emotions from strong agreement to disagreement or partial disagreement.

TABLE 12–3 Some Typical and Occasionally Inaccurate Statements about the ISO 9000 Series

Quotes	Author's Comments
"ISO 9000 ensures a company's systems are consistent. So, if you produce a poor quality product or service, ISO simply ensures you produce it consistently" (Tim Underhill, president of Underhill and Associates, ISO 9000 and TQM consultant).	This comment is often cited as an example of what is wrong with the ISO 9000 series. Unfortunately, it is inaccurate because if the company actually has evidence of poor quality, then it is expected to implement corrective actions and could not ship these products day in and day out. Several ISO 9001 or 9002 paragraphs (particularly 4.13 and 4.14) could actually be cited as evidence of nonconformance. Let us not forget that the market will sooner or later take care of companies that consistently deliver bad product.
"ISO 9000 got its beginnings from the first quality documented quality standards published by the military in the late 1940s in a document titled Mil Q9858" (Scott Stratman, "ISO 9000: How It All Began," p. 8.).	The reader should know by now that MIL-Q-9858 was first published in 1959 and not in the late 1940s. This is yet another example of a so-called expert spreading inaccurate rumors.
"ISO 9004 is an important tool which many companies don't use. But auditors look for it when they do an assessment. Many things that are only referenced in 9001 are more fully explained in 9004" (Chris Bohler, Qualitivity Inc.).	This is a partly correct statement. The 9004 tool certainly provides valuable *guidelines* often ignored by companies. There is, however, nothing wrong with ignoring 9004 because auditors are *not* supposed to refer to it during their audits simply because registered auditors will *never* audit a company using the 9004 guidelines. Another example of misinformation unfortunately spread via the media.
"There are two types of corporations: those who just want the certificate, and those who are going through the process because they are interested in improving their quality systems" (advertisement from an ISO 9001 certified company, p. 16).	The statement implies that one type of company is better than the other. Yet a company with an excellent quality record and excellent processes may only want to obtain the ISO 9000 certificate precisely because a customer demands it (an obvious example would include Motorola). A company that has improved its processes should not be assumed to be better than any other company. Indeed, the fact that it had to improve its processes prior to obtaining ISO 9000 certification may indicate that there was a lot of room for improvement!
"Because so many large U.S. firms are preparing for ISO certification in order to compete in international markets, many of	This is one of the classic statements linking certification with international competition. The two are certainly not

TABLE 12–3 *(Continued)*

Quotes	Author's Comments
their suppliers are beginning to feel the same pressure. . . . In other words, if you sell to a company that maintains an ISO 9000 system and your product or service can affect the quality of their product, that customer must 'qualify' you" (Bruce M. Kennedy, VP of Quality International, p. 19).	related except in the sense that, in some cases, a company may be required to achieve ISO 9000 registration to sell overseas. However, many companies with excellent quality records have had to achieve ISO 9000 registration. The requirement has little to do with quality improvement and is closer to a soft trade barrier. The second sentence is simply false. It is however true that some companies automatically require their suppliers to achieve ISO 9000 registration.
"U.S. manufacturers are currently playing catch-up in the world market and especially in ISO 9000 certification" (Kennedy, p. 19).	Another false statement. With more than 5500 (in 1994 when the quotation was written) sites registered, U.S. registration to ISO 9000 is ranked second in the world.
"While TQM is a philosophy, ISO 9000 is a road map—it's more concrete and offers a tremendous guideline for continuous improvement that isn't going to die" (Ray Gilles, quoted on p. 25).	Nothing wrong with the quotation except the optimistic certainty that the standard is immortal!
"ISO 9000 has also helped to keep our suppliers honest through the objective evidence we can now track—identifying errors and what it cost to correct those errors" (Joe Oven, p. 25).	Apparently, suppliers were not honest before the emergence of ISO 9000! Actually, ISO 9000 registration does not look into the costs of quality. Perhaps the individual is confused by the 9004 guidelines.
"Since quality control is a given with the ISO certification, Mailine's customers know they will receive the right product at the right time, at the correct price—over and over again" (Joe Oven, p. 25).	ISO 9000 certification cannot guarantee that a customer will receive the right product at the right time at the correct price. The pricing of a product is not addressed by the ISO 9000 standards. Many ISO 9000 registered firms still occasionally ship the wrong product.
"We want to work with fewer and more focused suppliers. . . . We want everyone to be on the same playing field. ISO certification is a way to accomplish both" (David Hertz, purchasing manager, p. 27).	The playing field analogy is often used. The quotation actually does reveal that some customers do use the ISO 9000 series as a means of eliminating suppliers (for the right or wrong reasons).
"Going through an ISO 9000 certification process, as Hetzer puts it, 'keeps you honest' in delivering what you say you will deliver" (David Hertz, p. 27).	This is an interesting quotation because it could imply that the supplier in question was actually dishonest prior to ISO 9000 registration. Apparently, prior to ISO 9000 registration, the supplier was *not* delivering what he had promised!

All quotations are taken from the special issue of *Today's Distributor*, June 1994.

TABLE 12–4 Chronology of the ISO 9000 Phenomenon as Reported in *Quality Progress* (1989–1992)

1989	First editorial by Nancy Karabatsos: "ASQC Moves Forward in Accrediting ISO 9000 Registrars," published in the July 1989 issue of *Quality Progress*.
October 1989	Ira J. Epstein (then with the secretary of defense and for a while vice president of Stat-A-Matrix, a well-known ISO 9000 consulting firm) writes a letter questioning the wisdom of Karabatsos' (i.e., ASQC) editorial and points out that ISO 9000 may be unfair to trade. Mr. Epstein has since change his mind considerably. In a May 1995 interview found in *Quality Systems Update*, regarding the DoD announcement about canceling MIL-Q-9858, we can read the following: " 'Now there's no doubt that contractors will be transitioning to ISO 9000,' explained Ira Epstein, a retired DoD officials who *championed* the acceptance of ISO 9000 in the Pentagon. 'This just puts the final nail in the MIL-Q-9858 coffin.' " To say that Mr. Epstein's letter of October 1989 was an example of championing the ISO 9000 series is stretching the imagination. Experts from Stat-A-Matrix are often quoted in *Quality Systems Update.*
January 1990	First ad for ISO 9000 by BQS: "ISO 9000 The European Community—1992." The ad sets the tone of fear that was to be repeated countless times by asking the reader: "Have you considered the potential impact on your European sales?"
May 1990	The same ad from January is repeated and states that the 12-nation community "will require compliance to their quality standards by year-end 1992." It asks: "What happens in 1992 if your company is not certified to the ISO standards?" The message continues; the need for certification is slowly being established.
June 1990	June 1990 issue of *Quality Progress* is devoted to the ISO 9000 standards.
June 1990	A letter from the United Kingdom points out that "ISO does little to ensure product quality."
October–December 1990	QMI states (ISO 9000) "Your Passport to Europe 1990." The reference to "passport" is later used by ABS and many others.
October 1991	Handley-Walker maintains that "You cannot compete in the European market without ISO 9000 certification." Also claims that ISO 9000 applies if you export to Europe, supply to European subsidiaries, etc. Handley-Walker ad states, "ISO 9000—Europe's Tougher Without It!"

TABLE 12–4 *(Continued)*

	Quality in Manufacturing prints its first ISO 9000 advertisement (BVQI).
1992	By 1992, a number of articles or editorials favoring sensationalistic titles begin to appear at an ever increasing rate. *Quality in Manufacturing,* for example, publishes: "ISO Standards Ensure Level Playing Field" (September–October 1992, p. 14). The expression "level playing field" is duplicated countless times. "ISO Update: Time to Panic?" (November–December 1992, p. 26). The article concludes "Not yet, but time is running out." Some of the articles are inaccurate or simply false. For example, the January–February 1993 issue has on p. 25 an article titled "ISO 9000 Ups Demand for Hardness Testing," in which the author tells the reader that the Brinell method has a slight edge but Rockwell is accepted. The truth is that the ISO 9000 standards make no reference whatsoever to Brinell or Rockwell hardness tests.

In late 1997, the first draft documents of the ISO 9001-2000 standard were being reviewed. The changes were significant and not everyone supported the new structure. The standards were reduced to one document (a good idea) but the 20-paragraph structure that had become familiar to tens of thousands of companies throughout the world were reorganized into 5 paragraphs. The standard was getting longer, going from 10 pages to about 12 or 13 pages. More requirements—referred to as clarifications by some committee members—were added and implementation as well as registration costs were likely to increase commensurably. The "more is better philosophy" was about to win once again. However, the important question that could not yet be answered was "Will all companies renew their certification to the new and more demanding ISO 9001-2000 standard?" Only time will tell.

Was ISO 9000 a Fad?

This paragraph was originally completed sometime in 1994 and was then entitled: "Is ISO 9000 a Fad?" Revisiting the paragraph 4 years later I found that some of the doubts I had back in 1994 were no longer doubts but certainties (thus the change from "Is" to "Was" in the heading). This is what I wrote in 1994:

The question is a legitimate one. Certainly, when one recalls all of the previous movements (as Juran would call them), skeptics may well ask: "Is ISO 9000 here to stay or will it be replaced by yet another fad/movement?" If we are to answer that question, one should first define fads. "A fad," Ken Hakuda tells us, "is something everyone wants yesterday and no one wants

TABLE 12–5 Analysis of Trends and Advertisements in *Quality Progress* (1990–1994, 1998)

Year	General Trends	Total Number of Ads and Promotions
1990	First ad in January issue	7
1991	Lead assessor courses appear. More ads stating that selling to Europe need not be a battle, etc.	75 ads avg. = 6.25 ads/month. Breakdown per month: 1 (4), 2 (3), 3 (4), 4 (4), 5 (7), 6 (4), 7 (5), 8 (9), 9 (9), 10 (7), 11 (6), 12 (11)
1992	Begin to see ads seeking auditors. Also begin to see more than one ad per page. Quality manuals, available on diskettes, make their first appearances for $97, later to drop to $75 only to rise again with more sophisticated programs selling for $300 to $700 or more.	158 avg. = 13 ads/month. Highest months: May and October with 22 and 23 ads, respectively.
1993	Begin to see BSi's ads for TickIT program. Begin to also see some consulting firms rely on multiple ads per issue. Some issues even have up to three ISO 9000 ads per page. New advertisers keep on coming in. The rush toward the ISO 9000 gold mine is still strong.	411 avg. = 34 ads/month. Highest month: May with 52 ads. Lowest month: January with 20. Estimated revenues for *QP*: $400,000 to $500,000 for ISO-related ads alone. Could multiply by at least a coefficient of 7 or 8 to include other periodicals such as *Quality Digest, Quality Magazine*, and dozens of other periodicals. Total could be about $3.5 to $4 million for the year.
1994	Ads seem to change. Fewer are for consultants or consulting and more registrars and software. Special ISO issue in May. Reference to ISO 14000 (environmental management system) makes its first appearance. Field seems more stable with fewer newcomers. Growth seems to have slowed down.	As of August, 286 ads or 36 ads/month. Estimate for the year about 432 or slightly higher than 1993. Stabilization! By the end of the year, the number of ISO 9000 ads dropped significantly only to be replaced by the latest fad: QS 9000 and eventually ISO 14000.
1998	The few ISO 9000 ads found in the July 1998 issue of *Quality Progress* advertise other services such as QS 9000 and ISO 14000. Business is definitely more difficult. Software designed to facilitate "your implementation task" is more prominent. Fancy software packages costing up to $2000 or more are now advertised.	Only 10 ISO 9000 ads with multiple messages. Is ISO 9000 dying? Will the trend pick up in 2000 when ISO 9001-2000 is published?

Note: The selection process was simple. Any ad that had the words "ISO 9000" or "Lead Assessor Training Course" was counted. The count does not include the want ad section or professional services section making up the last three or four pages of each issue.

tomorrow."[32] This definition would certainly seem to apply to the ISO 9000 movement. Unfortunately, despite some negative press and editorials such as Brad Stratton's "Good-Bye, ISO 9000; Welcome Back, Baldrige Award," found in the editorial comment of the August 1994 issue of *Quality Progress*, it is too early to tell if fewer and fewer companies will want ISO 9000 registration next year. In fact, the trend would indicate that ISO 9000 registration is probably here to stay for at least a few more years. Still if we are to believe ISO 9000 promoters and their advertisements, and if we recall the emergent skepticism voiced by a minority of observers, one could say that the ISO 9000 movement or phenomenon is on the verge of exhibiting the characteristics of a fad. What has occurred with the ISO 9000 phenomenon/fad is that the amount of misinformation, occasional disinformation, and rumors has been promoted for so long and with such intensity, that the public no longer knows what is right or wrong about ISO 9000.

Four years later I can assert what I already knew in 1994 but was unwilling to admit: ISO 9000 is no longer the "thing to do." This is not to say that more companies will not register to ISO 9000 in years to come, but that obtaining ISO 9000 certification no longer conveys the element of elitism it once had; it has become nothing more than a certificate of accomplishment that, in many cases, is now relatively easy to obtain (when you select the "right" registrar). Like all other fads before it, 9000 was a fad with a modest life cycle (in the United States) of about 6 to 7 years (SPC lasted much longer). Still, not bad for a fad and longer than most.

Is ISO 9000 a Legitimate Paradigm for the 21st Century?

"The survival of an idea," Tamotsu Shibutani tells us, "depends on its continued utility."[33] Certainly, if the ISO 9000 series were to add no utility to its users, it would surely disappear sooner or later. But is that necessarily true?

When we speak of utility, whose utility should we consider? The "utility" to customers, to suppliers/organizations, to registrars, to the training organizations, to the many ISO committees, to the International Organization for Standardization, to global society in general? Everyone perceived the utility of the 9000 series from a different perspective. Ultimately, one needs to consider the following question: "If, the standards were not perceived as a mandatory requirement to conduct business, would companies still seek ISO 9000 registration?" Unfortunately, I doubt that most companies would. We are therefore faced with yet another obvious question: "Do the ISO 9000 standards provide a legitimate paradigm for the 21st century?" I would suggest that it depends on their applications. I cannot be categorical in my answer because, based on testimonials pub-

---

[32]Ken Hakuda, *How to Create Your Own Fad and Make a Million Dollars*, William Morrow and Company, New York, 1988, p. 19. For additional studies on fads, see Peter L. Skolnik, *America's Crazes, Fevers & Fancies*, Thomas Y. Crowell Co., New York, 1972; Rene Konig, *A La Mode: On the Social Psychology of Fashion*, The Seabury Press, New York, 1973; and Marc-Alain Descamps, *Psychologie de la Mode*, Presses Universitaires de France, Paris, 1979.

[33]Tamotsu Shibutani, *Improvised News: A Sociological Study of Rumor*, The Bobbs-Merrill Company, New York, 1966, p. 182.

lished in various periodicals and my own experience, one cannot deny that achieving ISO 9000 registration has been of benefit to some (however, one does not know how long these benefits will last). Even in such cases, though, I would propose that the evidence must be carefully analyzed.

First of all, most, if not all, of the articles describing the multiple advantages achieved since registration are written by quality managers or directors who either wanted their companies to achieve ISO 9000 registration or were told to pursue registration. Moreover, these articles are written soon after the company has achieved ISO 9000 certification. But even if that assumption is incorrect, I would hardly expect a person to, after spending tens of thousands of dollars and numerous months of labor in order to achieve registration, confess that the effort was not worth it or that it had, at best, marginal benefits. Certainly, most everyone will readily admit that the effort and return on investment was worth it.

In addition, although developed nearly 50 years ago, many of the requirements outlined in the new ISO 9001-2000 standard, when implemented wisely and rationally, will be valuable to a host of manufacturers, assemblers, and service providers. By a rational application of ISO 9000, I mean:

1. Procedures must be developed with the *active* participation of employees-workers. Moreover, management *and* employees should periodically review procedures to constantly improve them. (Ironically this approach would suggest a partial return to Frederick Taylor's misunderstood scientific management.)[34]
2. An emphasis on continuous training (an unending task).
3. The development of a practical, simple, and easy-to-use documented system.

Yet, if we recognize that achieving ISO 9000 registration can be beneficial to manufacturers (in the traditional sense of the word), will it guarantee world-class competitiveness, as some have argued? The answer must be categorically "no." It cannot be so for at least two obvious reasons:

1. One cannot achieve world-class status by merely registering its quality assurance system to an international standard. To claim that this is possible is simply illogical because standards, by their very definition, can only provide minimum requirements. Also, let us not forget that if everyone is registered, can we then assume that "everyone" has achieved world-class competitiveness? Of course not.

---

[34]Juravich observed that workers want to create quality products but to do so, workers must be allowed more decision making over issues that directly affect their work tasks (Juravich, *Chaos on the Shop Floor*, p. 148). Juravich also explains that American management has misapplied Taylor's ideas by focusing solely on the management aspect of scientific management and ignoring the need to analyze and document the skills of workers (i.e., "craft knowledge") to develop new principles. These new principles would then be used to train workers. See Juravich, *Chaos on the Shop Floor*, pp. 87–90, and Frederick Winslow Taylor, *The Principle of Scientific Management*, W. W. Norton, New York, 1911, p. 36.

2. The standards were derived from military standards. This is not necessarily intrinsically bad, but we must remember that these military standards were developed for a very unique customer–supplier relationship that does not always apply in the private sector except for a few cases such as the automotive industry and regulated industries (Marquardt's multilevel versus generic standards). Moreover, such contractual standards are not cost effective and therefore cannot be, nor should they be, applied to a scenario subject to the basic rules of profitability.

The same objections would apply to the upcoming revision, ISO 9001-2000. If anything, the ISO 9001-2000 standard is likely to complicate rather than simplify life for organizations: more procedures, more documents, more surveys, more measurements, more design control, more paperwork, more, more. Experts are diametrically opposed as to what companies will need to do to survive: Tom Peters believes that companies will need to "thrive on chaos"; the ISO TC 176 committee and its supporters believe that the key to success resides within the pages of the next ISO 9000 revisions.[35] Who do you think is right? Will chaos reign, which would help foster flexibility and quick response, or is regimented, controlled order prescribed by an international standard better? I suspect neither extreme is the answer. How to balance chaos with some form of regimented control may well be the challenge for the next century!

What has remained unexplained to this point is how and why the ISO 9000 series achieved such popularity within a relatively short period of time. The next chapter attempts to answer this question.

---

[35]Tom Peters, *Thriving on Chaos*, Harper Perennial, New York, 1987.

CHAPTER 13

Quality Professionalism and the Ideology of Control

When a large scale activity has developed giving a good livelihood to a group of people, they will use all kinds of propaganda to prove it is in the public interest and morally right.

—The Law of Vested Interest (as stated by Professor Meredith Thring of Suffolk, England)

. . . being sold solutions is better than being confronted with questions about problems.

Neil Postman[1]

Instead of attempting to discover what is most significant with the highest degree of precision possible under the existing circumstances, one tends to be content to attribute importance to what is measurable merely because it happens to be measurable.

—Karl Mannheim[2]

What Is Professionalism?

In his *Culture of Narcissism*, the sociologist Christopher Lasch writes that when professionalism emerged in the 19th and early 20th centuries, it was not as a result of well-defined social needs. "Instead," Lasch observes, "the new professions themselves invented many of the needs they claimed to satisfy."[3] Lasch's comments certainly seem to be true when one reviews the historical evolution of the medical, engineering, and quality professions.[4] When these invented needs are challenged or questioned by society, professional organizations are quick to

---

[1]Neil Postman, *Amusing Ourselves to Death*, Penguin Books, New York, 1986, p. 131.

[2]Karl Mannheim, *Ideology and Utopia*, Harcourt, Brace & World, New York, 1967, p. 85.

[3]Christopher Lasch, *The Culture of Narcissism*, W. W. Norton & Company, New York, 1979, p. 385.

[4]For an excellent description of the medical profession, see Paul Starr, *The Social Transformation of American Medicine*, Basic Books, New York, 1982. The history of the engineering profession is told by Edwin Layton, Jr., *The Revolt of the Engineer: Social Responsibility and the American Engineering Profession*, Johns Hopkins University Press, Baltimore, 1986.

defend themselves. For example, when the California legislature proposed a bill (AB 969) that would eliminate the need for quality engineers to be registered, the American Society for Quality (ASQ)—which played a major role in creating and defining the profession of quality engineers and which earns a substantial income from certifying quality engineers in the United States and abroad—felt threatened and immediately reacted by opposing the bill.[5]

California's AB 969 also indicates that the California legislature apparently agrees with Daniel Bell's assessment of professionalism. Indeed, some years ago, Bell wrote: "[T]he effort to 'professionalize' work has become the major means of giving one's job a badge of honorific quality which the nature of the work itself denies."[6] Could it be that California's AB 969 is an attempt to demonstrate that the honorific title of quality engineer bestowed by the ASQ is not as valuable as the professional society would have us believe?

Professionalization, Magali Sarfatti Larson contends, is a process by which producers of special services seek "to constitute *and control* a market for their expertise"[7] [emphasis added]. To be recognized as a professional, Ms. Larson proposes that the professionals must be adequately trained and socialized so as to provide recognizably distinct services for exchange on the professional market. Once this is achieved, unification must follow. This is most easily achieved when a group of professionals is "ready to champion the propagation of one 'paradigm,'" and when the group has "enough persuasive or coercive power to carry the task through."[8] Although the willingness to champion one and only one paradigm may be too restrictive a criterion, the fact remains that one of the reasons why the quality profession might have had difficulties being seriously accepted as a profession by others is that it often treats paradigms as if they were fads, recycling them every 3 to 5 years, and sometimes posing one against the other (Malcolm Baldrige Award versus ISO 9000 registration, for example).

One could also argue that quality professionals have not yet achieved complete recognition because as quality engineers, they are mostly perceived as technicians, not always aware of or conscious of the business requirements of a company. This problem of "marginality" has certainly been experienced by the engineering profession in general. As Edwin Layton explains, the engineer is expected to be both a scientist and a businessman, but in fact he is neither and thus, although his technical knowledge is appreciated, he is still perceived as marginal to the business.[9] Professionalism also carries overtones of ideological elitism that may alienate those outside the profession. But what is meant by an ideology of quality? Is it an immutable concept?

---

[5]"California's Engineer Registration May Cease," ON Q, July–August 1998, p. 9. The ASQ not only administers several certification tests but also offers many seminars and offers a broad spectrum of courses relating to quality engineering.

[6]Daniel Bell, *The End of Ideology*, Harvard University Press, Cambridge, 1988, p. 257. Originally published in 1960.

[7]Magali Sarfatti Larson, *The Rise of Professionalism*, University of California Press, Los Angeles, 1977, p. xvi.

[8]Ibid., p. 40.

[9]Edwin T. Layton, Jr., *The Revolt of the Engineers*, Johns Hopkins University Press, Baltimore, 1986, Chap. 1, "The Engineer and Business," pp. 1–24.

The Ideology of Quality

Karl Mannheim noted some years ago,

The concept of "ideology" reflects the one discovery which emerged from political conflict, namely, that ruling groups can in their thinking become so intensively interest-bound to a situation that they are simply no longer able to see certain facts which would undermine their sense of domination. There is implicit in the word "ideology" the insight that in certain situations the collective unconscious of certain groups obscures the real condition of society both to itself and to others and thereby stabilizes it.[10]

Mannheim's observation regarding the concept of ideology applies rather well to the quality establishment, particularly when one considers the continued struggle experienced during the past 30 or so years by practitioners of the various disciplines of quality to be recognized as professionals as opposed to mere technicians.[11]

One of the difficulties faced by all professions is that they tend to become isolated from other communities. This phenomenon, repeatedly observed by sociologists, slowly evolves as professionals begin to limit their interactions with other professionals who share their views. This group-think effect had been noted in studies performed in the 1940s by the American sociologist Paul Lazarfeld who wrote that "[T]hose who read the press of their group and listen to the radio of their group are constantly reinforced in their allegiance. They learn more and more that their group is right, that its actions are justified; thus their beliefs are strengthened."[12] As the anthropologist Edward T. Hall explained many years ago: "The individuals of a group share patterns that enable them to see the same thing and this holds them together."[13] This reinforcement of ideas, concepts, beliefs, and principles occurs at least once a year when thousands of quality professionals gather at the ASQ quality congress to exchange and/or confirm various paradigms: statistically inclined practitioners attend statistically

---

[10]Karl Mannheim, *Ideology and Utopia*, Harvest, New York, 1967, p. 135.

[11]Surprisingly, prior to 1970, relatively few articles had been written about the quality profession. The handful of papers published within the pages of the *ASQC Transactions* and in *Industrial Quality Control* from 1960 to 1965 generally focused on defining academic requirements for the quality engineer. Approximately 40 articles have been published by *Quality Progress* from 1968 to 1992. For early articles, see W. H. Lewis, "Progress in Professionalism," in *ASQC Convention Transactions*, ASQC, San Francisco, CA, 1960, pp. 303–306; William R. Collins and Samuel S. J. Skolnik, "The Evolution of an Occupation," *Industrial Quality Control*, April 1962, pp. 4–5; K. C. Asay and J. W. Morris, "Professionalism in Quality Control," in *ASQC Annual Technical Conference Transactions*, ASQC, Los Angeles, CA, 1965, pp. 50–56; Steve Kozich, "Professionalism," in *ASQC 1965 Transactions*, ASQC, Los Angeles, CA, pp. 120–135; and Donald D. Stewart, "Quality Control: An Emergent Profession," *Industrial Quality Control*, July 1965, pp. 9–11.

[12]Lazarfeld quoted in Jacques Ellul, *Propaganda: The Formation of Men's Attitude*, Alfred A. Knopf, New York, 1965, p. 213.

[13]Edward T. Hall, *The Silent Language*, A Fawcett Premier Book, Greenwich, Connecticut, 1959, p. 115. Dr. Hall's other contributions include *The Hidden Dimension*, Anchor Books, Garden City, New York, 1966; *The Dance of Life*, Anchor Books, New York, 1983; *An Anthropology of Everyday Life*, Anchor Books, New York, 1993.

oriented lectures, auditors attend audit sessions, and so on through the various subdisciplines.

The belief system and thus the ideologies (for there are many) of the profession are further reinforced with the evolution of a quality jargon that helps further alienate quality practitioners from fellow corporate members of the corporation. This alienation exhibited by the classic "us-versus-them" syndrome, is witnessed daily in numerous companies and can be explained, in part, by what linguists refer to as the Whorfian hypothesis. Benjamin Whorf's hypothesis from the 1920s postulates that "the structure of a human being's language influences the manner in which he understands reality and behaves with respect to it."[14] Whorf contended that because man dissects nature along lines laid down by his native language, reality becomes relative and therefore "all observers are not led by the same physical evidence to the same picture of the universe, unless their linguistic backgrounds are similar, or can in some way be calibrated."[15] These observations would certainly apply to the quality profession (or any other profession for that matter). Indeed, although quality professionals do speak the same language as their coworkers, they do not share the same quality vocabulary (and, hence, ideology), rich in expressions, acronyms, or abbreviations such as JIT, DOE, SPC, CQE, CQA, AQL, TQM, TQC, CWQA, QA, QC (control and circles), Cpk, benchmarking, Baldrige, ISO 9000, etc., which can easily confuse the noninitiated. This linguistic jargon of quality-speak, developed by quality professionals over the years, has certainly helped quality practitioners establish themselves as a professional group (something it could not have achieved without the continued and dedicated efforts of the American Society for Quality). However, this process has resulted in two divergent side effects. On the one hand, quality professionals have somewhat succeeded in differentiating themselves from others. On the other hand, the process has also facilitated the standardization of ideas neatly categorized into topics or tracks and ultimately simplified and transformed into a myth beyond recognition by the use of slogans.

This is precisely what Jacques Ellul proposes in his study on propaganda. To Ellul, ". . . ideologies emerge where doctrines are degraded and vulgarized and when an element of belief enters into them."[16] The element of belief, or blind faith, is important because it allows followers to degrade, vulgarize, and simplify concepts (as some alarmist environmentalists are notoriously known for). These tasks, Ellul contends, are achieved by propagandists who then take over and transform the ideology into myths. The advantage of propaganda, Ellul notes, is that it furnishes man

> . . . *with a complete system for explaining the world, and provides immediate incentives to action. We are in the presence of an organized myth that tries to take hold of the entire person. Through the myth it creates, propaganda imposes a complete range of intuitive knowledge, susceptible of only one*

---

[14]Benjamin Lee Whorf, *Language, Thought, and Reality*, The M.I.T. Press, Cambridge, MA, 1973, p. 214.

[15]Ibid., p. 214.

[16]Ellul, *Propaganda*, p. 12.

interpretation, unique and one-sided, and precluding any divergence. This myth becomes so powerful that it invades every area of consciousness, leaving no faculty or motivation intact. . . . Propaganda cannot be satisfied with partial successes, for it does not tolerate discussion; by its very nature, it excludes contradiction and discussion.[17]

Within the quality profession, the mythology of quality has led to a vulgarization of quality concepts and ideologies exemplified daily by the endless and often mindless repetition of trite slogans and one-liners. A specific example of the very issues raised by Ellul in the preceding quote will help illustrate the point. If we are to believe a survey published by Grant Thornton in 1994, 79 percent of companies surveyed are aware of the international models for quality assurance known as the ISO 9000 series.[18] Based on my own informal survey conducted over the years while flying from one assignment to the next, I would propose that although some people are aware of the ISO 9000 series of standards, most (i.e., more than 60 to 70 percent) have never heard of the ISO 9000 series of standards.

I have selected the ISO 9000 series because it is the most recent example of a worldwide "quality phenomenon" which would satisfy Ellul's definition of propaganda. During the late 1990s, I noticed that whenever I attempted to raise some issues or dared to question the validity of some of the claims made by proponents of the ISO 9000 movement, quality professionals and consultants alike often relied on the same patented answers. These answers invariably revolved around the theme of "It's good for us (that is, America)." However, when I ask my friends what metrics they have used to measure this "goodness" or "improvement factor," most would confess that they have no such metric yet, they simply feel that "it" is better than before. This typical position is probably best exemplified with the following anecdote.

As we were debating the virtues of the 9000 series a consultant friend of mine began to passionately propose a defense of the standards based on socialistic principles by suggesting that the ISO 9000 series summarized the collective wisdom of quality experts. As such, my friend contended, the series provided a service to society as a whole because it required businesses to finally recognize their responsibility and debt to society and the environment as a whole (he was including the ISO 14000 series of environmental standards). When I tried to counter by asking my friend to explain how a set of standards that has its roots in military specifications could represent the voice of the (millions of) customers and suppliers, the answer drifted back to classic arguments of the type: "It is a good solid fundamental system which will help American companies become more competitive." How could one argue with goodness and increased competitiveness? The validity of my friend's statement is of course questionable and, besides, has nothing to do with the original question.

---

[17]Ellul, *Propaganda*, p. 193. For a description of some contemporary myths, see Roland Barthes, *Mythologies*, Hill and Wand, New York, 1998; see particularly his essay "The Blue Guide," pp. 74–77.

[18]"ISO 9000 Is Not Dead!" *Quality Progress*, November 1994, p. 23.

The ISO 9000 ideology, Ellul would contend, is slowly being transformed into the ISO 9000 myth. But how has the myth evolved? Surely, the ideology of quality did not grow out of nothing; it evolved and developed over several decades; it has a history.

Influence of the Military and the Ideology of Quality Control

To understand why and how the quality movement became dominated by the military establishment during the 1940s and through the 1980s, it is necessary to first understand how, until the early 1990s, the military has permeated just about every aspect of the socioeconomic life of most Americans. Soon after World War II, the future of the military–corporation alliance that had helped win the war was sealed when in 1946, General Eisenhower published an influential memorandum in which he outlined the purpose and objectives of the "military industrial complex."[19] The memorandum outlined five basic principles:

1. The Army must have civilian assistance in military planning as well as for the production of weapons.
2. Scientists and industrialists must be given the greatest possible freedom to carry out their research.
3. The possibility of utilizing some of our industrial and technological resources as organic parts of our military structure in time of emergency should be carefully examined.
4. Within the Army we must separate responsibility for research and development from the functions of procurements, purchase, storage, and distribution.
5. Officers of all arms and services must become fully aware of the advantages which the Army can derive from the close integration of civilian talent with military plans and developments.[20]

The principles outlined by Eisenhower were to have a profound impact on American socioeconomic life, including the field of quality control and quality assurance.

By 1955, when the military budget averaged more than 51 percent of the entire federal budget, the sociologist C. Wright Mills observed that "As the role of the government in the economy has increased, so has the role of the military in the government."[21] This phenomenon, Mills observed, led to the formation of a new power elite in America. This power elite emerged during the mid-1950s, as retiring generals and admirals began to penetrate corporate America. The number of corporations that had retired generals on their board is rather impressive. In 1955, Mills wrote that the following corporations had generals on their board:

---

[19]Eisenhower's memo is reproduced in Appendix A of Seymour Melman's *Pentagon Capitalism*, McGraw-Hill Book Company, New York, 1970, pp. 231–234.

[20]Eisenhower quoted in Melman, *Pentagon Capitalism*, pp. 232–233.

[21]C. Wright Mills, *The Power Elite*, Oxford University Press, New York, 1956, p. 212.

- Continental Can Company
- Shell Oil
- Bulova Research Laboratories
- Remington Rand (Douglas MacArthur, chairman of the board)
- AVCO Corporation
- Jones & Laughlin Steel Corp.
- Fairchild Aircraft Corp.
- Hughes Tool Co.
- Koppers Co.
- Mercast Inc.
- Remington Rand (General Leslie R. Groves, vice president)
- Lockheed Aircraft Corporation
- American Machine Foundry Company
- Kaiser's automotive division
- Mellon Institute of Industrial Research

The appointment of retired military personnel was not limited to corporate headquarters. Indeed, Mills notes that science/research and educational institutions were also favorite targets. "The elite," Mills concluded, "cannot be truly thought of as men who are merely doing their duty. They are the ones who determine their duty, as well as the duties of those beneath them. They are not merely following orders: they give the orders. They are not merely 'bureaucrats': they command bureaucracies."[22]

This slow but steady influx of military men within the private sector had a direct influence on the quality profession and on its ideology.

Influence of the Military on the Perception of Quality

The importance of the military during the formative years of the American quality movement is most clearly demonstrated by reviewing early issues of *Industrial Quality Control* (IQC). The periodical, founded in 1944, was published bimonthly by the Society of Quality Control Engineers in cooperation with the University of Buffalo (New York). Reviewing the 23 years of publication (IQC ceased publication in 1967), one is struck by the dominance of articles published by university professors, military organizations, or subcontractors to the military. Table 13–1 reveals that most of the articles, if we include contributions from organizations such as General Electric and other subcontractors with substantial military dependency, are contributions from military organizations or military

---

[22]Ibid., p. 286. The following universities had admirals or generals as chancellors, deans, or members of the board of trustees: Lewis College of Science and Technology, University of California at Berkeley, Ithaca College, George Washington University Law School, St. Louis School of Medicine, California State Board of Education (ibid., p. 219). For a critical review of Mills' argument, see Daniel Bell's "Is There a Ruling Class in America? The Power Elite Reconsidered," in Daniel Bell, *The End of Ideology*, Harvard University Press, Cambridge, MA, 1988, pp. 47–74. For a brief criticism of Bell as a labor sociologist, see Harry Braverman's *Labor and Monopoly Capital*, Monthly Review Press, New York, 1974, pp. 106–107.

TABLE 13–1 IQC Number of Publications by Organization and by Periods (Ranking in Parentheses)

Institution or Organization	1944–1950	1951–1960	1961–1967	Overall Ranking (in parentheses)
University*	30 (1)	56 (1)	30 (2)	116 (1)
Military	06 (2)	25 (2)	42 (1)	73 (2)
General Electric	04 (5)	20 (4)	14 (3)	38 (3)
Bell	06 (2)	22 (3)	07 (8)	35 (4)
Carborundum	—	08	14 (3)	22 (5)
Sylvania	03	05	10 (5)	18 (6)
Westinghouse	04 (5)	04	07	15 (7)
IBM	—	05	09 (6)	13 (8)
Sandia	—	01	12 (4)	13 (8)
N.B. Standards	—	11 (5)	01	12 (9)
Kodak**	03	04	04	11 (10)
Martin	—	02	08 (7)	10 (11)

*University: Purdue with I. Burr has more than a dozen contributions (probably 14 or 15 during the 30-year period [11 by the end of 1960]).

**Kodak or Eastman Kodak (also from various divisions).

Note: Examples of what I refer to as articles published by military sources would include the following: U.S. Naval Propellant, AFB Utah, Norton AFB Reliability, Bureau of Naval Weapons Reliability/Missiles, Department of the Navy, Defense Supply and Logistics, Military Clothing and Textile, Army Chemical Center, Defense Production, Naval Weapon, Frankfort Arsenal, Ballistics Systems, Secretary of Defense, Department of the Navy, USAF Material Evaluation, Wright-Patterson AFB, U.S. Navy Commander Naval Air Force, U.S. Army Mobility, U.S. Naval Ordnance, Ballistic Research Lab., Chemical Warfare Service, Office of Chief Ordnance, Chemical Corps Engineering Army Chemical, Guided Missile, Ordnance Ammunition, U.S. Naval Engineering, Office Chief Ordnance, U.S. Naval Powder Factory, etc.

Other important contributors: Western Electric (7), Raytheon (6), General Dynamics (9), Gillette Safety Razor (6), Avco (6).

Interesting mention in the 1950s: articles by the IRS, AFL, and CIO.

subsidized organizations. Indeed, if we add to the 73 contributions directly made by military personnel, the contributions of some obvious subcontractors to the military (i.e., GE, Bell, Westinghouse, Raytheon, Martin, Western Electric, General Dynamics, Avco), the count conservatively jumps to about 350 to 370 articles out of a total of 833 for the 23-year period.

When one reviews the content of the articles from Table 13–1, a pattern emerges. First, one notices that over the 23-year period, as many as 32 to 39 percent of the articles were published by no more than six organizations (Table 13–2). This concentration is revealing when one analyzes the contents of the articles, which, until the early 1960s, dealt almost exclusively with inspection sampling or statistical quality control in general.

Reviewing the events that occurred during the formative years of the American quality control movement, one soon understands how and why the military came to favor the use of statistical techniques. The story begins in 1936 when Walter Shewhart, generally acclaimed as the father of control charting, served as

TABLE 13–2 Measure of Concentration

Year	Number of Articles and Percentages
1944–1950	47 articles or 35 percent published by four "organizations"
1951–1960	123 articles or 32 percent published by four "organizations"
1961–1967	124 articles or 39 percent published by six "organizations"

a consultant on ammunition specifications for the War Department. As a result of Shewhart's activities, quality control methods were introduced at Picatinny Arsenal.[23] In May 1941, under Shewhart's leadership, the Z War Standards were produced. The publication of these standards is significant because they became the foundation of a course in quality control offered in 1942 by Eugene L. Grant of Stanford University. By 1943, the course had been expanded to an 8-day course known as the ESMWT (Engineering, Science, and Management War Training). Although I have been unable to review the contents of these courses, it would appear that the emphasis was on statistics rather than management. Indeed, there is little doubt that even though few authors wrote about the management of quality control, the dominant theme of the 1950s was statistical control. In fact, as far as some authors were concerned, a person's quality control IQ was directly related to one's knowledge of statistics.[24] This limited perception of quality by a few professionals indicates the particular prejudices of the time. Consequently, from the mid-1940s on, major corporations were essentially dictating and propagating a specific ideology regarding the nature of quality control.[25]

---

[23]The following account is based on several articles: Paul S. Olmstead, "Walter Andrew Shewhart," *Industrial Quality Control*, February 1956, pp. 5–6; Martin A. Brumbaugh, "Highlights in the History of the ASQC," *Industrial Quality Control*, February 1956, pp. 6–11; Eugene L. Grant, "Look Back and Look Ahead," *Industrial Quality Control*, July 1953, pp. 31–35; Colonel Leslie E. Grant, "The Advancing Frontier of Quality Control," *Industrial Quality Control*, May 1949, pp. 5–9; and Leslie E. Simon, "Quality Control at Picatinny Arsenal, 1934–1945," *ASQC Technical Conference Transactions*, ASQC, Chicago, IL, May 19–21, 1971, pp. 273–279.

[24]See, for example, R. O. Swalm and C. R. Hicks, "What's Your IQ in QC?," *Industrial Quality Control*, October 1959, pp. 84–85. Out of 30 questions, 26 relate to one's knowledge of statistics. The author scored 16 out of 30, which is labeled "Good" and apparently good enough (in 1959) to consider senior membership in the society.

[25]This is not to say that other topics and points of view were not presented for indeed they were; but rarely. In fact, many of these earlier issues provide the reader with some very valuable historical accounts relating to the evolution of the American Society for Quality Control. Still, besides these occasional historical accounts, what is interesting to observe during that period is that deviations from the main train of thoughts seemed to have come from essentially one voice, namely, J. M. Juran. I am referring to the string of articles published by Dr. Juran in *Industrial Quality Control* from the 1946 to 1967 where he provides the reader with some refreshing thoughts. See, for example, his "The Relation of the Quality Function to the Industrial Enterprise," *Industrial Quality Control*, July 1946, pp. 5–7; "Installing a Quality Control System," *Industrial Quality Control*, May 1953, pp. 21–23; "Goals

To these men, quality meant quality as practiced by the leading companies of the time. And since the leading companies of the time were likely to be contractors to the military, the quality establishment of the time merely duplicated quality requirements imposed by the government. Because the government was a customer, then surely, what was good for the government must have been good for others. Hence, from 1944 to 1967, quality professionals were bombarded by articles constantly reminding them of the virtues of statistical process control (SPC), inspection (required by various military specifications), supplier surveillance (MIL-Q-5923 and later MIL-Q-9858), and, from 1962 on, total quality control as expounded by Dr. Armand Feigenbaum of General Electric.

The popularity of these techniques called for by the growing number of military standards soon began to spill over into the private sector. At Ford Motor Company, for example, the use of statistical process control became so popular that a supervisor (not a manager) was assigned the responsibility of the Chart Section. Besides the extensive use of SPC techniques, Ford implemented other techniques favored by the military. PERT charts, which were used everywhere at Wright-Patterson Air Force Base, also became popular at Ford despite the enormous overhead cost.[26]

This transfer of quality principles and ideology from the government to the private sector, which began during the 1940s and 1950s, continued for another two to three decades. The promulgation of the principles of quality control as dictated by the government were dutifully promoted by various leaders of the quality community. As Table 13–3 reveals, these leaders came from major corporations and often were prime military contractors. (*Note:* No small firms, defined as firms with 50 or fewer employees, were ever represented.)

Having had to implement or work with quality control and quality assurance systems designed to satisfy MIL-Q-5923 and later MIL-Q-9858, it was logical for these "experts" to write about what they knew.

Reviewing the themes of the various conferences and transaction papers published from 1947 to 1960, one is struck by the concentration of topics, the uniformity of the message, and the communality of themes found in IQC (see Table 13–4). Representatives from the armed services, occasionally relying on patriotic arguments, constantly justified the value of military specifications and supplier surveillance as a means to ensure that America's defense would remain strong.

---

for Quality Management in the Next Decade," *Industrial Quality Control*, May 1962, pp. 9–15; "Pioneering in Quality Control," *Industrial Quality Control*, September 1962, pp. 12–14; "The Two Worlds of Quality Control," *Industrial Quality Control*, November 1964, pp. 238–244 (where Juran warns of the danger of focusing only on the statistical techniques of quality control); "Quality Control under Monopoly," *Industrial Quality Control*, March 1965, pp. 462–464; "Quality Problems, Remedies and Nostrums," *Industrial Quality Control*, June 1966, pp. 647–653 (where Juran criticizes the zero-defect movement); and "A Challenge to the Extinction of 'Homo Inspectiens,'" *Industrial Quality Control*, December 1967, pp. 298–299 (where Juran offers perhaps one of the earliest criticisms of the internal inspector).

[26]W. M. H. Smith and C. R. Burdick, "Ford's Interest in Statistical Quality Control," *Industrial Quality Control*, November 1950, pp. 6–14, and Michael McGill, *American Business and the Quick Fix*, McGraw-Hill, New York, 1988, pp. 11–12.

TABLE 13–3 The Quality Establishment as Represented by IQC-ASQ(C) Presidents for the Years 1945–1995

Year	President	Company
1945	Alfred J. Winterhalter	Colonial Radio Corp.
1946–1948	George E. Edwards (New York)	Bell Telephone Labs.
1949	Ralph E. Wareham (New York)	National Photocolor Corp.
1950–1951	Wade R. Weaver	Republic Steel Corp. Cleveland
1952	Simon Collier	Johns-Manville Corp.
1953	Alfred L. Davis	Rochester Institute of Technology
1954	Raymond S. Saddoris	A. O. Smith Corp.
1955	Paul A. Robert	IBM
1955	Arthur Bendler	Delco-Remy GMC
1956	Dale Lobsinger	United Air Lines
1957	Leon Bass	General Electric
1958–1960	C. E. Fisher	Bell Telephone
1960–1961	J. Y. McClure	Convair General Dynamics
1961–1962	Armand Feigenbaum	General Electric
1962–1963	J. Y. McClure	General Dynamics
1963–1964	E. Jack Lancaster (Wayne, NJ)	American Machine & Foundry Co.
1964–1966	Rocco L. Fiaschetti (Downey, CA)	North American Aviation, Inc. Space and Info. System Div.
1966–1967	William A. Golomski (Milwaukee, WI)	Joseph Schlitz Brewing Co.
1967–1968	Robert M. Berg (South Charleston, WV)	Union Carbide Corp.
1968–1969	Thomas C. McDermott (Downey, CA)	North American Rockwell, Space Division
1969–1970	Thomas E. Turner (Indianapolis, IN)	Diamond Chain Co.
1970–1971	L. I. Medlock (San Diego)	Convair Div. General Dynamics
1971–1972	David S. Chambers (Knoxville, TN)	University of Tennessee
1972–1973	Richard A. Freund (Rochester, NY)	Eastman Kodak, Kodak Park
1973–1974	Harry J. Lessig (Springfield, IL)	Stewart-Warner Corp., John W. Hobbs Div.
1974–1975	Howard L. Stier (Boston, MA)	United Brands Co.
1975–1976	Charles H. Brokaw (Atlanta, GA)	Coca Cola USA
1976–1977	Orde R. Weaver (Bartlesville, OK)	Phillips Petroleum Co.
1977–1978	Walter L. Hurd (Burbank, CA)	Lockheed Aircraft Corp.
1977–1978	John S. Dermond (St. Petersburg, FL)	Honeywell, Inc.
1978–1979	Robert N. Reece (Hunt Valley, MD)	McCormick & Company, Inc.
1979–1980	Philip B. Crosby (New York)	International Telephone & Telegraph Corp.

TABLE 13–3 *(Continued)*

Year	President	Company
1980–1981	Jay W. Leek (Winter Park, FL)	The Quality College, Inc.
1981–1987*		
1987–1988	J. Douglas Ekings	Xerox Corp.
1988–1989	Spencer Hutchens, Jr.	Intertek Services Corp.
1989–1990	John E. Condon,	Abbott Laboratories, TRW
1990–1991	John Knappenberger	
1991–1992	Charles A. Aubrey II	Banc One Corp.
1992–1993	Robert V. Caine	G.E. Aerospace
1993–1994	Charles A. Aubrey II,	Juran Institute Inc.
	Robert V. Caine	Chairman of the Board G.E. Aerospace
1994–1995	Jack West	Westinghouse Electric Corp.

*1981–1987 data not collected.

TABLE 13–4 Breakdown of Dominant Themes as Reflected in the ASQC Conference and Transaction Papers (1947–1960; number of articles = 911)

Topic/Theme	Number of Articles
Armed services	53
Missiles	2
Standards	20
Specifications	10
	Total = 85 or 9.33%
Statistics	
Analysis of variance	13
Binomial distribution	1
Chi-square	3
Control charts	40
Correlation/regression	16
Design of experiment	27
Reliability	28
Sampling	57
Statistics	35
	Total = 220 or 24.0%
Inspection and testing	60
Vendor relationship	22
Visual characteristics	7
	Total = 89 or 9.76%
Metals	67 or 7.3%
Management	45 or 5.0%
Automotive	27 or only 2.9%

Proponents of statistical techniques never ceased to demonstrate the virtues of statistical techniques, and quality control inspectors tirelessly proposed that inspection would save the customer money.

Of course, when these authors spoke of the customer, they invariably were referring to the federal government as a customer. To them, there was no distinction between an omnipresent customer and the millions of other customers-consumers. E. G. D. Patterson typified this attitude when he stated:

> *In reviewing some of the pertinent literature in order to get a feeling for the general attitude toward the inspector, I reread an article "Government and the Inspector". . . . I cannot recall any article better expressing, from the consumer's standpoint, the need for the plain, old-fashioned inspector, without change in title, organization, or purpose.*[27]

Recognizing that the article in question applied to military procurement, Patterson concluded by stating that, "its arguments and conclusions apply equally well to non-military industry."[28] This argument, repeated by others, is fundamentally flawed for several reasons. First of all, Patterson fails to distinguish between a customer with unlimited resources and therefore economic clout (the government) or very powerful customers such as GM, Ford, and Boeing and customers that must operate within well-defined market constraints and budget requirements and which are therefore limited in their ability to impose, or even define, one-way requirements.

Second, Patterson wrongly assumes that the economic burden associated with the implementation of a quality assurance system designed to satisfy government regulations is independent of the size of the company. In other words, Patterson repeats the fundamental mistake made by quality consultants and management consultants during the past few decades by assuming that the business constraints of small firms are identical to those of very large firms. Yet, as John A. Welsh and Jerry F. White have so clearly demonstrated, "the very size of small businesses creates a social condition—which can be referred to as *resource poverty*—that distinguishes them from their larger counterparts and requires some very different management approaches."[29] Consequently, the facile assumption, repeated in various styles and in countless papers that what is good for big business is necessarily good for small businesses is fundamentally wrong. As Walsh and White concluded:

> *Owner-management of a small business is a distinct discipline characterized by severe constraints on financial resources, a lack of trained personnel, and a short-range management perspective imposed by a volatile competitive environment. Liquidity must be a prime objective. The analytical models applica-*

---

[27]E. G. D. Patterson, "Stop, Look, Inspect," *Industrial Quality Control*, August 1964, pp. 88–89.

[28]Ibid., pp. 88–89. For a counterargument, see J. M. Juran, "A Challenge to the Extinction of Homo Inspectiens," *Industrial Quality Control*, December 1967, pp. 288–299.

[29]John A. Welsh and Jerry F. White, "A Small Business Is Not a Little Big Business," *Harvard Business Review*, July–August 1981, pp. 18, 32.

ble to big business are of limited use in this arena. Typically, they assume steady-state conditions subject to minor changes.[30]

Finally, Patterson does not seem to recognize that whenever large businesses want quick turnaround they invariably subcontract the work to hundreds of small workshops whose managerial *flexibility*, unhindered by government regulations and inspectors, and ability to communicate better, usually can guarantee rapid turnaround.[31] Oddly enough, very few quality practitioners have written about this. My review of the quality literature, as represented by such periodicals as *Industrial Quality Control* and *Quality Progress*, revealed only one article on the subject. In an article published in 1963 and entitled "Quality Control Management of Small Business," Henry Jacobson points out that since small companies (which he generously defined to be organizations with fewer than 300 employees) do not have the communication problems of their larger competitors, they can operate with far smaller staffs. "This factor of communication," the author noted, "is one that gives the edge to the smaller firm. I have always maintained that no large corporation could compete pricewise with a small efficient company making the same product."[32]

This ideology of supplier (quality) control, modeled after various military quality requirements and propagated by numerous quality professionals from 1944 to this day, is further exemplified by a series of documents published by the American Society for Quality.

The Role of the ASQ(C) in Promoting Supplier Regulations

During the 1970s through the 1990s, the American Society for Quality Control—which was renamed in 1997 as the American Society for Quality (ASQ)—published several documents whose intent was to provide guidelines on how to better "control" suppliers.[33] I will not review the contents of these publications, but will instead examine who the contributors were and what major influences and/or sources of information were used to produce the document. Table 13–5 summarizes findings for three such documents.

Perusing Table 13–5, one notices that, predictably, contributors to standards are more likely to come from major corporations, the very type that can afford to

---

[30]Ibid., pp. 18, 32.

[31]Christy Borth (*Masters of Production*, The Bobbs-Merrill Company, New York, 1945, p. 175) recognized back in 1945 that "[I]t was from such little shops, serving the larger factories with specialty tools in peacetime but capable of transforming themselves into suppliers of precision parts for assembly into finished weapons elsewhere, that the automotive industry was able to derive the aid needed to produce fire power in quantity and in time."

[32]Henry J. Jacobson, "Quality Control Management of Small Business," *Industrial Quality Control*, March 1963, p. 5.

[33]There is perhaps a strange irony when one considers the name change of the American Society for Quality (ex-control), because although the society has removed the word "Control" in its name, the trend in the field of quality in general has been toward more control. This is achieved via the publication of seemingly unending international standards.

TABLE 13–5 Sample Review of Contributors to ANSI/ASQC Standards

American National Standard, *Generic Guidelines for Auditing of Quality Systems* (sponsored by the ASQC's Quality Auditing Technical Committee), January 1986.

No members listed; however, the appendices list the following references (partial listing):

1. Canadian Standards Association.
2. *ANSI/ASME Qualification of Quality Assurance Program Audit Personnel for Nuclear Power Plants,* N 45.2.23 and document N.45.12.12, *Requirements for Auditing of Quality Assurance Programs for Nuclear Power Plants.*
3. ASQC Energy Division, *Nuclear Quality Systems Auditor Training Handbook*, ASQC, Milwaukee, 1980.
4. U.S. General Accounting Office, *Standards for Audit of Governmental Organization, Programs, Activities and Functions*, 1972.
5. Delroy L. Cornick, *Auditing in the Electronic Environment*, Lomond Books, Mt. Airy, MA, 1981.

How to Establish Effective Quality Control for the Small Supplier, Vendor–Vendee Technical Committee, ASQC, Milwaukee, 1985.

The 13 authors came from the following companies/professions: Naval Systems Command, Western Electric Co., Fortune Ind., Wang Laboratories, Argyle Associates (a consulting group), Telecommunications & Electronics Branch, Transport Canada, Lorillard (a division of Loew Theaters!), Babcock & Wilcox, Bendix Corp., Didak Corp., and Lockheed Georgia Co.

The technical committee consisted of some 42 members representing industries such as Fiat-Allis North America, Inc., General Electric Co., Fokker, B.V., Digital Equipment Corp., 3M Center, Martin Marietta Energy Systems, Westinghouse Electric Corp., Abbott Laboratories, and Northern Telecom Canada Ltd.

How to Evaluate a Supplier's Product, Vendor–Vendee Technical Committee, ASQC, Milwaukee, 1981.

The authors came from the following companies: Fiat-Allis Construction Machinery, Action Communication Systems, Newport News Shipbuilding, Burndy Corp., General Electric Co., Bendix Corp., and Transport Canada.

subsidize the individual's participation. Consequently, although the committee may be called the vendor–vendee technical committee, the dominant influence is from the particular customer–vendee point of view of the likes of GE, Naval Systems Command, Westinghouse, Fiat, Bendix, Abbott Laboratories, and so on. One wonders if the vendors' points of view are ever represented. When committee members are not listed, one notices that the tendency for these committees is to automatically rely on previous works created for very different scenarios.

The auditing of quality systems documents, for example, relies heavily on the auditing practices found in the nuclear industry or those practiced by the U.S. government. There is, of course, nothing wrong with reviewing what others have done because this should always be a requirement; however, one should refrain from simply adopting recommendations made for one industry where the societal risks of any failure are obviously high or where the customer (government) has unusual powers. Systems developed for the auditing of nuclear plants have an inherent cost structure that is commensurate with their application. Consequently, although the system may be suitable for the unique customer–supplier environment for which it was designed, it is often inadequate or, worse yet, irrelevant to the needs of others. Yet, rather than reconsider how a new system should be developed (assuming a system is even needed), the committee members invariably end up mimicking their references without fully realizing the consequences or impact their actions might have when the system is later interpreted by others (regulating agencies, for example). Identical errors are being committed by the committee in charge of the ISO 9001-2000 revisions.

Recent Trends: The Ideology of Management (Soft Quality)

In looking at Table 13–6, from 1968 (the year *Quality Progress* was first published) to 1992, one notices that the range of subjects published is significantly broader with an obvious bias toward management issues (200 out of 1822 articles).

Each reader will probably interpret the significance of the numbers given in Table 13–6 differently. From my perspective, I find it particularly interesting that the subject of Continuous Improvement, so dear to quality professionals, was only addressed twice in 25 years. Customer Satisfaction, yet another fundamental axiom of quality, was only written about 10 times, and Participative Management 6 times. (Is this because it is obvious or not worth writing about?) Small Business was only mentioned twice (should we be surprised?) and Warranty only once! I find it also interesting to note that, although there are plenty of articles on Quality Costs, apparently few of these articles deal with the economics of quality assurance and/or quality control from the overall company perspective. This inability on the part of quality professionals to associate quality with the overall economy of the firm was observed over 30 years ago by J. M. Juran.

The Quality Function and the Economy of the Firm

J. M. Juran seems to have been the only quality professional in the past 30 or so years to have analyzed the schism and dichotomy between quality professionals and management. As early as 1964, Juran recognized that there were two worlds of quality control: the world of top management, which is primarily concerned with performance of the quality function; and the world of the quality specialists, who are primarily concerned with the specialties within the quality function—that is, the techniques of quality.[34] Juran suggested that in recent years

---

[34]J. M. Juran, "The Two Worlds of Quality Control," *Industrial Quality Control*, November 1964, pp. 238–244. Juran defines the quality function as the means through which a company discovers and meets the quality needs of its customers.

TABLE 13–6 *Quality Progress* Subject Index and Count (1968–1992; number of articles = 1822)*

AQC Address (12)	Edwards Medal (4)	Military (15)	Quality System Accreditation (4)
ASQC Survey (13) Interviews (8)	Electronics (5) Energy (21)	Motivation (8) NASA (3)	Reliability (30) Research and Development (6)
Auditing (31) Automotive (12)	**Environment (60)** Equipment (48)	NASA Award (16) National Quality Month (40)	Robotics (1) Sampling (32)
Benchmarking (11)	Food, Drug, and Cosmetics (14)	Nondestructive Testing (10)	Self-Improvement (1)
Certification (5)	Government (8)	Participative Management (6)	Service Quality Assurance (34)
Communicating Quality (14)	Grant Award (3)	Performing Arts (3)	Shewhart Medal Acceptance (9)
Communication (7)	Graphical Analysis (3)	Planning for Quality (5)	Small Business (3)
Community Quality Improvement (1)	Health Care (39)	Problem Solving (1)	**Society (68)**
Computers (26)	History (11)	Process Capability (3)	Software (23)
Configuration Assurance (7)	Human Resources (17)	Process Control (8)	Special Report (38)
Conflict Resolution (3)	Inspection (48)	Process Improvement (54)	Standards (52)
Consumerism (10)	International (62)	Procurement (1)	Statistical Process Control (14)
Collector's Series (11)	Just-In-Time (5)	Product Development (21)	Statistics (45)
Continuous Improvement (2)	Lab. Accreditation (1)	Product Liability (48)	Systems (12)
Control Charts (3)	Malcolm Baldrige Award (23)	Productivity (5)	Systems Engineering (8)
Culture (2)	**Management (200)**	Profile: Honorary Members (13)	Taguchi Methods (10)
Customer Satisfaction (10)	Manufacturing (1)	Quality Circles (18)	Teamwork (2)
Customer–Supplier (49)	Manuf. Systems (2)	Quality Costs (56)	Technology (20)
Data QA (10)	Marketing (9)	Quality Forum (22)	Testing (19)
Decision Making (2)	Measurement (3)	Quality Function Deployment (9)	Tools of Quality (11)
Deming Prize (2)	In Memoriam (4)	Quality Manuals (1)	Total Quality (7)
Design of Exp. (2)	Metals (2)	Quality Profession (41)	Training (33) Variation (14)
Education (34)	Metrology (19)	Quality Support Groups (6)	Warranty (1) Zero Defects (3)

*The top three categories are shown in bold characters.

Source: Quality Progress 25-Year Cumulative Index. 1968–1992.

(he was referring to the 1960s but the observation is still valid today), quality activities shifted from the broad quality objectives of meeting the quality needs of its customers, "to the much narrower objective of meeting the quality specification,"[35] which naturally led quality practitioners and management to emphasize and focus on technical problems and technique rather than business problems. By 1976, Juran had sharpened his concept when he correctly observed that "making product which is defect-free does not mean the product is surely saleable."[36] Thus, to management, the interest in quality is driven by the fact that a quality product "is essential to securing the company's income." Yet, Juran observed, amidst the plethora of papers on the cost of quality, essentially none addressed the issue of the effect of quality on income! Juran suggested that this was predictable because quality specialists were not sufficiently business oriented. Since most quality specialists were trained as engineers, "they have tackled the problems with which they are familiar (and) they have stayed out of the 'business' problem of income."[37]

Since Juran published his paper in 1976, quality management issues have certainly not been ignored. One could even suggest that the pendulum has now shifted almost exclusively on the side of management. Yet despite this emphasis on the management of quality, Juran's observations are still valid today. Quality practitioners still seem to be totally oblivious to the economic constraints of the firm and the need to maintain income. But because quality specialists are rarely involved with business decisions affecting income, one should not blame them for a company's failure.

Conclusion

One of the problems with professionals and/or experts is that people outside the clique tend to confuse intellectual merit with professional credentials "or worse, with loyalty to an unspoken ideological consensus."[38] Describing how citizens have fallen prey to the "illusory fantasies" of professional bureaucrats, Ludwig von Mises explains:

> *In order to break the existing pattern of dependence and put an end to the erosion of competence, citizens will have to take the solution of their problems into their own hands. They will have to create their own "communities of competence." Only then will the productive capacities of modern capitalism, together with the scientific knowledge that now serves it, come to serve the interests of humanity instead.*[39]

---

[35]Ibid., p. 240.

[36]This and the following quotation from Juran are taken from his "The Quality–Profit Relationship," in *ASQC Technical Conference Transactions*, ASQC, Toronto, Canada, 1976, pp. 18–29. The software industry would prove Juran right (see Chapter 5).

[37]Juran, "The Two Worlds of Quality Control," p. 22.

[38]Lasch, *The Culture of Narcissism*, p. 386.

[39]Von Mises is quoted in Lasch, *The Culture of Narcissism*, p. 396.

If industrialists and business owners in general want to eliminate or at least reduce their dependency on the opinions and ideology of professional experts, they will need to follow Mises' suggestion. To do this, business owners must first try to counter Professor Thring's Law of Vested Interest by reasserting their own (business) common sense. The sooner the business community can do that, the sooner they can get down to the business of conducting *their* business.

Part V

Consequences of Standardization

CHAPTER 14

On the Origin of Procedures

Experience is a wonderful thing. It enables you to recognize a mistake when you make it again.

> —Words of wisdom written on the back of a sugar pack found
> in a coffee shop in a small town somewhere in Utah

Thou shall not write down thy principles, still less print them, lest thou shall be entrapped by them for all time.

> —Neil Postman[1]

I do not know who wrote the first procedure or when it was first written. Was it written in Mesopotamia by an Assyrian, or in China by a bureaucrat working for one of the early Ming dynasties, or perhaps in ancient Egypt by one of the pharaoh's physicians? One would like to know if, among the thousands of clay tablets and papyrus discovered by anthropologists throughout the world during the past several decades, any described an assembly or "manufacturing" procedure. This would be an interesting question to ask anthropologists.

Naturally, one must recognize that people have long relied on verbal explanation and demonstration of how a task is performed simply because few could write and fewer still could read. Still, one wonders if Greek and Roman engineers followed (written) procedures when they built their great temples, aqueducts, roads, or other engineering works that still stand to this day. Did ancient Egyptians build the pyramids following plans or were they relying on knowledge, perhaps secret, that was passed verbally to an elite few?[2]

Perhaps procedures were not written down until relatively recently because the written word was not always trusted. Zuboff writes:

The medieval historian M. T. Clanchy has illustrated the reluctant acceptance of written documentation in place of first-person witness as it occurred over

---

[1]Neil Postman, *Amusing Ourselves to Death*, Penguin Books, New York, 1986, p. 31.

[2]Juran notes that specifications written on papyrus scrolls were found to be more than 3500 years old. But no mention is made of procedures, the word is not even listed in the index. J. M. Juran, *A History of Managing for Quality*, ASQC, Milwaukee, 1995, p. 607.

more than three centuries of English history. "Documents," he tells us, "did not immediately inspire trust." People had to be persuaded that written documentation was a reliable reflection of concrete, observable events. . . . To members of a highly oral culture, however, the spoken word was connected to the incontrovertible reality of bodily experience, while the written word was a thin, substanceless scratching whose two-dimensionality seemed highly arbitrary.[3]

It was not until the early part of the Middle Ages that the production of documents began to increase. Ivan Illich and Barry Sanders note that as royal mandates began to increase from the period 1080–1180—from 3 to 60 for French kings, 25 to 115 for English kings, and 22 to 180 for popes—"The consumption of sealing wax at the royal chancery in England rose from three pounds per week in 1226, to thirteen pounds in 1256, and thirty-one pounds just ten years later in 1266. More sheep had to give up their skins as parchments for the purposes of documentation during a royal hearing."[4]

As more and more documents were being produced, Illich and Sanders observe that the meaning and function of a signature began to take effect for legal purposes in the 13th and 14th centuries. Documents were still read in the 12th century, but by the 13th and 14th centuries the signature was increasingly important particularly as a written warranty (from the old French *warandir*).[5]

It would thus appear that at least Western European documents were first written to codify the edicts produced by nascent bureaucracies that emerged throughout the European medieval world. One would have to wait until the 17th or 18th century to begin to see early attempts in the Western world at "proceduralizing" processes—often in the form of pictures accompanied by text. In France, for example, the school of Encyclopedists initiated by Diderot (1713–1784) began to document thousands of processes in the textile and other industries.[6]

Procedures during the Dawn of Industrialization

The events that were to lead early American industrialists to recognize the importance of procedures can be traced to at least the mid-18th century and the

---

[3]Shoshana Zuboff, *In the Age of the Smart Machine*, Basic Books, New York, 1988, p. 77. It seems to me that medieval thoughts were ahead of us because written documents are rarely followed to the letter.

[4]Ivan Illich and Barry Sanders, *The Alphabetization of the Popular Mind*, North Point Press, San Francisco, 1988, p. 36.

[5]Ibid., p. 43. Also the business of making exact copies took off in the 13th century when paid correctors began to check/verify documents "according to form (*ratio*), legibility (*lettera*), and spelling (*sillibo*)." The dating of documents also became important (ibid., pp. 40, 41).

[6]For examples, see Dominique Chevalier, Pierre Chevalier, and Pascal-Francois Bertrand, *Les Tapisseries D'Aubusson et de Felletin (1457–1791)*, Solange Thierry Editeur, La Bibliotheque des Arts, 1988. Sidney Pollard in his *The Genesis of Modern Management. A Study of the Industrial Revolution in Great Britain*, Harvard University Press, Cambridge, MA, 1965, states that work flow was often planned in the iron industry by the early 1700s. Pollard also cites a book of instructions published in 1742 that mentions work flow and planning (p. 263).

then-emerging need for the manufacture of interchangeable military parts. To satisfy the demand of the French army for interchangeable standardized parts, General Jean-Baptiste Gribeauval developed in 1765 what became known as "le systeme Gribeauval."

By the early 1780s the American army had recognized the importance of the Gribeauval system and by 1798 the U.S. War Department was issuing contracts specifying interchangeable parts with private arms makers. Simon North and later Eli Whitney received such a contract for 20,000 muskets and pistols. The age of mass production and the American system of production had arrived. In 1824, John T. Hall successfully manufactured 1000 rifles with completely interchangeable parts.

Despite the eventual success of these early manufacturers, several decades went by before Henry Leland—who had worked as a tool builder for the U.S. Armory at Springfield, Massachusetts, in 1863—began to institute in the 1880s at the Browne & Sharpe sewing plant, the procedural approach to manufacture. As David Hounshell explains:

All operations on various parts were enumerated on sheets along with the necessary tools, jigs, fixtures, and gauges. . . . Leland's insistence that these operations be recorded offers a commentary on the mechanic who had discovered the "Art of Manufacture" and who had taken an intense interest in process rather than the building of any particular production tool.[7]

By 1886, Singer had abandoned the European handwork mode of production—which favored employing cheap workmen to finish each piece—in favor of its blue book approach "which delineated all of the machining operations and work-flow routes."[8] The blue book, Hounshell writes, "codified Singer factory production operations for the first time in the company's history."[9]

By the turn of the century, the demands of mass production were being recognized by an increasing number of manufacturers in the United States and abroad. Despite the recognition, only a few manufacturers such as the Ford Motor Company began to implement procedures throughout their plants to try to control process repeatability and prevent uncontrolled process changes.

Heritage of the American System

The American system of production and its continuous search for perfectly interchangeable parts has been critically reviewed by some historians. David F. Noble sees flaws in the American system of production and explains that "the overriding impulse behind the development of the American system of manufacture was military; the principal promoter of the new methods was not

---

[7]David Hounshell, *From the American System of Mass Production 1800–1932*, The Johns Hopkins University Press, Baltimore, 1985, p. 82.

[8]Ibid., p. 82.

[9]Ibid., p. 120.

the self-adjusting market but the extra-market U.S. Army Ordnance Department." Noble sees the development of interchangeable-parts manufacture as an "expensive hobby" of a "customer with unlimited funds and was dictated by military criteria of uniformity and performance, regardless of cost."[10] To make matters worse, Noble also notes that the American system, which finds its roots in military contractual requirements, "was subsequently encouraged and carried over into civilian production (agricultural implements, sewing machines, bicycles) by arsenal personnel who brought with them a military enthusiasm for uniformity and automaticity that reinforced the growing industrial obsession, epitomized by Andrew Ure, with 'perfecting' production by eliminating labor."[11]

Another unfortunate side effect of military contracts—which was to be rediscovered countless times during the next 100 or so years—was that the demand for precision, often imposed by military contracts, forced subcontractors to the military to implement costly organizational structure for receiving, in-process, and final inspection of parts and processes as well as burdensome document control and record retention procedures. As Hounshell astutely observes, "[T]he bureaucratic structure that evolved in American arms production to make and maintain the many precision gauges that ensured uniformity was absent in clockmaking."[12] Unfortunately, the success of the ISO 9000 series of standards, which is itself derived from old military standards, has forced the private sector to duplicate many of the costly requirements first imposed in the United States by the military back in the early 1780s. Examples of costly practices would include elaborate procedures for instrument calibration and precision control as well as tedious and sometimes complex document control mechanisms for all documented procedures.[13]

Procedures: Anathema, Panacea, or Placebo

If we are to believe some individuals, especially most quality consultants and fans of the ISO 9000 series of standards, no organization can operate without procedures. To these individuals, the development of procedures is a panacea that will relieve any business of any and all problems. Yet, based on the past 10 years of personal experience, I would have to say that many successful businesses, particularly successful small businesses, have either few or no procedures or perhaps, worse yet, obsolete procedures. Of course, some busi-

---

[10]David F. Noble, *Forces of Production: A Social History of Industrial Automation*, Oxford University Press, New York, 1986, p. 334.

[11]Ibid., p. 334. The French sociologist Jacques Ellul says the same. See also Harry Bravermann, *Labor and Monopoly Capital*, Monthly Review Press, New York, 1974.

[12]Hounshell, *From the American System*, p. 56.

[13]As Noble (*Forces of Production*, p. 339) points out: "As we have seen, the chief impulses behind the development of N/C—like those which apparently spurred the emergence of the American system of manufacture a century earlier—were not simply economic. Rather, they reflected the combined and compounded compulsions, interests, beliefs, and aspirations of the military, management, and technical enthusiasts, as they do today" and "computerization requires standardized procedures" (ibid., p. 343).

nesses do maintain up-to-date procedures, but, to my knowledge, no research has ever been conducted to demonstrate conclusively, preferably with statistical data, that well-documented or "proceduralized" businesses are more successful or have fewer customer complaints than businesses with no or few procedures. Could it be that procedures often act much as a placebo would for certain patients?

For many people, the writing and existence of procedures convey a general spirit of evil. When "management consultant Jack Gibb once asked a group of managers to design an organization that would produce the lowest levels of trust among its employees, some of the answers were: Make sure everything is locked up, install time clocks, introduce voluminous manuals of operating procedures, develop rules and regulations on everything."[14]

To others, such as author-consultant Peter Block and the late sociologist Harry Braverman, the documentation of procedures is nothing more than a managerial form of control over workers. Block, for example, proposes that it is bureaucracy that demands consistency and thus procedures.[15] In an age of industrial technology that has reduced the labor process and the importance of work and human talent, less skill (but not necessarily less training) is now required in many entry-level jobs. Workers, Shoshana Zuboff notes, have been *reduced to following* procedures.[16] At Sony, in Tijuana, Mexico, for example, assemblers and testers are nothing more than human appendages to the verification software (computerized procedures) designed to do all the "thinking and testing." Green lights and red lights tell the assembler where to adjust magnets to guarantee a perfect image. Yet, despite such objections, the age of procedures and procedure writing (accelerated in the past decade by the ISO 9000 phenomenon) does not appear to exhibit any signs of weakening.

On Working Knowledge

Little has been written about procedures, their effectiveness and value to employees or workers, and the assembly process in general. As early as 1922, George S. Radford, in what is perhaps the first book on quality published in the United States, devoted a few pages to what he refers to as the "written description of a process." Radford recognized the difference between

> *... knowing the method and principles used in doing the work, on the one hand (the why and how) and the skill required for their execution, on the other. One could write out the most particular instructions for shooting a rifle, but would only acquire the skill necessary for accurate shooting through contin-*

---

[14]Quoted in Richard Farson's *The Management of the Absurd*, Simon and Schuster, New York, 1996, p. 131. Even renowned science writer Stephen Jay Gould admits to a general dislike for procedures and standardization and homogenization in general (Stephen Jay Gould, *Eight Little Piggies. Reflections in Natural History*, W. W. Norton & Co., New York, 1993).

[15]Peter Block, *Stewardship*, Berrett-Koehler Publishers, San Francisco, 1993, p. 26. See also Braverman, *Labor and Monopoly Capital.*

[16]Zuboff, *In the Age of the Smart Machine*, p. 23.

uous practice. Yet almost anyone could learn to shoot by following the written instructions exactly.[17]

Very much influenced by Taylor's scientific method, Radford optimistically believed that Taylor's scientific method could be used to describe any job. As far as Radford was concerned, "such a description may be used as the basis for improvement, once the governing principles have been worked out; and it can be employed as well to start any other intelligent person toward acquiring the skill needed in its execution."[18]

More than half a century later the sociologist Ken Kusterer was to elaborate further on the subject of "working knowledge." Kusterer recognizes two types of knowledge associated with work: basic and supplementary (which he later refers to as "working knowledge"). "Basic knowledge," Kusterer explains, "includes all the procedures necessary to routinely carry out their (workers) work tasks. . . . Supplementary knowledge includes all the know-how necessary to handle the obstacles to this routine work performance that arise from time to time: how to keep the machine running, overcome 'bad' paper, diagnose the cause of defects, keep the inspectors happy. . . ."[19] Kusterer later explains that "When all the conditions are right, and nothing in the work environment interferes, this basic level of knowledge about procedures is all that is necessary to accomplish the assigned work tasks in a highly productive way."[20]

Naturally, environmental conditions are not always "right." Variability often interferes with these ideal conditions and this is when the working knowledge comes in to solve all of these problematic conditions not foreseen or addressed during basic training (or idealized work conditions).

What is particularly interesting about Kusterer's book is that his research complements much of what I observed as a management consultant during the 1990s. For example, Kusterer recognizes that since it is with working knowledge that a worker learns the idiosyncrasies of her or his machine, working knowledge is essential to help a worker resolve problems (i.e., find the causes of problems); a skill usually not addressed by the formal basic training offered by companies.[21] One also must recognize that, with working knowledge, workers can often go directly to the cause (source) of the problem—that is, the cause directly responsible for product defect(s). There is often no need for time-consuming cause-and-effect diagrams and analysis generated by cross-functional teams. Yet, in many cases, management, who is likely to have invested a substantial amount of money in training and/or consulting fees, would sometime insist that workers do

---

[17]George S. Radford, *The Control of Quality in Manufacturing*, Ronald Press Company, New York, 1922, p. 388.

[18]Ibid.

[19]Ken C. Kusterer, *Know-How on the Job: The Important Working Knowledge of "Unskilled" Workers*, Westview Press, Boulder, CO, 1978, p. 45. Kusterer's working knowledge is equivalent to Tom Juravich's craft knowledge and similar to Frederick Taylor's "traditional knowledge." See Tom Juravich, *Chaos on the Shop Floor*, Temple University Press, Philadelphia, 1985.

[20]Kusterer, *Know-How on the Job*, p. 139.

[21]Ibid., pp. 49, 51, 166.

not use their "intuition" (i.e., working knowledge) to solve problems. This failure to recognize the value and latent asset of working knowledge and only favor the official well-documented company methodologies often leads to elaborate and inefficient solutions; hence the success of Scott Adams' famed cartoon series known as *Dilbert*.[22]

Another important aspect of working knowledge is that it helps workers reduce or eliminate the feeling of alienation by fostering a spirit of community. Workers, Kusterer explains, with their working knowledge of processes, learn to develop *a communal network of relationships* associated with their jobs. This communal network (referred to by sociologists as the *Gemeinschaft* as opposed to the *Gesselschaft* of management) is built by workers to bring meaning to their work. This phenomenon, well documented in Kusterer's book, is important to recognize. Indeed, when managers want to bring about changes, they do not realize that their actions are likely to simultaneously destroy or, at a minimum, threaten the communal network of relationships and built-in meanings that workers have worked hard to build over the years. Managers do not understand the resistance to change often presented by workers who see their communal network threatened.[23]

The next chapter explores further the subject of procedures: why they are written and when and how they should be written.

---

[22]I would also point out that it is this type of working knowledge that is required to conduct effective statistical studies such as design of experiments (DoEs), for example.

[23]Kusterer, *Know-How on the Job*, p. 158.

CHAPTER 15

Writing Procedures

Introduction

When I purchased a ping-pong table, I was glad to find included with the table a set of instructions and (poor) diagrams explaining how to assemble the table. Although I hardly referred to the instructions (which did cause me to err on one occasion), I did find the diagrams, poor as they were, helpful. There is no doubt that assembly instructions are useful for first-time assemblers. But what about people who assemble ping-pong tables for a living? Would they refer to the instructions and/or diagrams every time they assembled a table? Obviously not. In cases where the assembly is rather simple, assemblers would probably ignore the instructions by the time they get to the second assembly.

Such is the interesting paradox of assembly procedures; once written, they soon become of little value to the "experienced" assembler who may need to refer to them only once or twice. Of course, if a new table is designed, or if new or different parts are introduced by the manufacturing or design engineer(s), a new set of instructions will need to be released to warn and inform the assembler of the changes. And yet, even in such cases, practical experience demonstrates that even without a set of new instructions, *experienced assemblers* very quickly notice the change and immediately adapt and modify their assembly method, often before a new set of instructions is formally released by manufacturing or engineering. Yet, one must also acknowledge that some assemblers never notice the design changes or are never told of the design changes and keep assembling according to the obsolete method, thus creating havoc.

Assemblers or workers in general are able to rapidly adapt to such changes either because the assembly process is relatively simple or because they are so familiar with the process they can almost instantly adapt to the changes. Unfortunately, one must also acknowledge that assemblers have been known to arbitrarily modify assembly instructions and simply ignore certain steps just because they feel it is more efficient or expedient to do so. One could argue that not all processes are as simple as the assembly of a ping-pong table, which, of course, is true. Still, many manufacturing processes or subprocesses have been so "modularized" as to be actually simpler (and more monotonous) than the assembly of a ping-pong table.

So why are procedures written? Procedures or more detailed work instructions are written to describe how a series of steps is to be performed. Why is this important? Procedures are written for a variety of reasons: to explain and/or demonstrate how a sequence of events is to be performed (e.g., ping-pong table assembly); to preserve knowledge (e.g., assembly knowledge); to teach new employees; and, if consistency is important, to ensure repeatability of a process.

Repeatability can be translated into a statistical concept of minimum or, ideally, zero variability. Although repeatability is desirable in many mass production manufacturing processes where tolerances need to be maintained and duplicated over hundreds if not tens of thousands of parts, process consistency-repeatability can be annoying and may well have to be avoided in certain processes. In recent years thousands of companies have purchased automated answering systems. Although there is no question that these answering systems offer reliable repeatability and monotony, it is arguable as to whether or not they offer customer satisfaction. I often find these predictably efficient answering systems annoying and not always efficient (from the customer point of view, that is). Repeatable or standardized processes are not likely to be a desirable feature in the service sector. This assumption is often erroneously posited by experts who unfortunately manage to convince companies to adopt standardized procedures to handle customer complaints; these can often lead, as we shall see later, to a ludicrous implementation of quality concepts.

Frederick Winslow Taylor on Procedures

In 1911, Frederick W. Taylor published his influential paper titled "The Principles of Scientific Management." The paper, based on extensive studies he conducted in the late 1870s and early 1880s at the Midvale Steel Company and during the mid-1890s at the Bethlehem Steel Company, was Taylor's attempt to answer President Roosevelt's "larger question of national efficiency." Taylor's answer to the many inefficiencies he saw surrounding him was to be *scientific management*. To Taylor, scientific management or task management, consisted of four underlying principles:

> *First. The development of a true science. Second. The scientific selection of the workman. Third. His scientific education and development. Fourth. Intimate friendly cooperation between the management and the men.*[1]

Taylor was convinced that the philosophical principle of scientific management would allow management to collect all of the information found in the *traditional knowledge* acquired over the years by workmen and "reduce this knowledge to rules, laws, and formulas which are immensely helpful to the workmen in doing their daily work."[2] Once management had studied, documented, and improved

---

[1]Frederick Winslow Taylor, *The Principles of Scientific Management*, Dover Publications, Mineola, New York, 1998, p. 68.

[2]Ibid., p. 15.

on the workers' traditional knowledge, it was management's responsibility to ensure that all workers were trained to these new methods. This would in turn guarantee that the workers' performance would improve, which would in turn greatly please management and eventually customers who would supposedly be able to obtain cheaper goods.[3]

Under certain conditions such as a machine-shop, for example, Taylor believed that detailed work instructions that described the best way to do a job needed to be prepared in advance by the planning department. Once these best methods were developed and became a standard, they could be superseded by yet another "quicker and better series of movements."[4] Workers, Taylor proposed, were to be encouraged to participate in this process of continuous improvement. Taylor concluded, "It should be the policy of the management to make a careful analysis of the new method, and if necessary conduct a series of experiments to determine accurately the relative merit of the new suggestion and of the old standard."[5] Moreover, Taylor emphasized that "The workman should be given the full credit for the improvement, and should be paid a cash premium as a reward of his ingenuity."[6]

Yet, Taylor was a practical man, and after more than 20 years of experience trying to convince businesspeople of the virtues of scientific management over the more traditional initiative and incentive managerial method, Taylor knew that change was not easily brought about and would take time to implement—Taylor suggested that as much as 5 years would be required to fully implement scientific management. Halfway through his paper, Taylor strongly emphasizes that "It is only through *enforced* standardization of methods, *enforced* adoption of the best implements and working conditions, and *enforced* cooperation that this faster work can be assured. And the duty of enforcing the adoption of standards and of enforcing this cooperation rests with the *management* alone."[7] Management, Taylor continued, must supply the teachers to show each man the new method and "The *management* must also recognize the broad fact that workmen will not submit to this more rigid standardization and will not work extra hard, unless they receive extra pay for doing it."[8] It is no wonder that many employees to this day still perceive standard operating procedures as nothing more than a management tool used to monitor and control their activities.

---

[3]Taylor (ibid., p. 71) cites how a Pennsylvanian Dutchman named Schmidt was taught how to improve his output by a factor of 4 while at the same time improving his pay by 60 percent. "It does seem grossly unjust," Taylor observes, "when the bare statement is made that the competent pig-iron handler, for instance, who has been so trained that he piles 3.6 times as much iron as the incompetent man formerly did, should receive an increase of only 60 percent, in wages. . . . At first glance we see only two parties to the transaction, the workmen and the employers. We overlook the third great party, the whole people—the consumers, who buy the product of the first two and who ultimately pay both the wages of the workmen and the profits of the employers." Taylor sees the right of the customer as being greater than the other two.

[4]Ibid., pp. 61, 64.

[5]Ibid., p. 67.

[6]Ibid.

[7]Ibid., pp. 41–42.

[8]Ibid., pp. 41–42.

Herbert Simon on Decisions

Herbert Simon distinguishes between two types of decisions: programmed routine, repetitive decisions for which organizations develop specific processes for handling them, and nonprogrammed one-shot, ill-structured, novel policy decisions which are handled by general problem-solving processes.

The techniques used to handle each decision type vary: Programmed decisions are traditionally related to (1) habits, (2) clerical routines, or (3) organizational structures. Modern approaches to programmed decisions would include (1) operational research or (2) electronic data processing.

Nonprogrammed decisions have been traditionally handled by (1) judgment, intuition, and creativity; (2) rules of thumb; or (3) selection and training of executives. The modern approach to nonprogrammed decisions favors heuristic problem-solving techniques (i.e., by experimentation).

Simon observes:

Habit is the most general, the most pervasive, of all techniques for making programmed decisions. The collective memories of organization members are vast encyclopedias of factual knowledge, habitual skills, and operating procedures. Closely related to habits are standard operating procedures. The only difference between habits and standard operating procedures is that the former have become internalized—recorded in the central nervous system— while the latter begin as formal, written, recorded programs. Standard operating procedures provide a means for indoctrinating new members into the habitual patterns of organizational behavior, a means for reminding old members of patterns that are used so infrequently that they never become completely habitual, and a means for bringing habitual patterns out into the open where they can be examined, modified and improved.[9]

An important side effect of programmed activities, which is particularly important to note for enterprises that have achieved ISO 9000 certification, is what is known as the Gresham's law of planning. The law states that programmed activities tend to drive out nonprogrammed activities. "The organizational implication of Gresham's Law," Simon warns, "is that special provision must be made for nonprogrammed decision making by creating specific organizational responsibilities and organizational units to take care of it."[10] Thus, one of the dangerous consequences of implementing standardized procedures is that a business may become "stale" and, as a result, become outdated. No wonder that the so-called

---

[9]Herbert Simon, *The New Science of Management Decision*, Prentice Hall, Englewood Cliffs, NJ, 1977, p. 50. Simon also notes that most organizations are faced with a large cost associated with hiring new employees who must learn the proper habits they need to perform at their jobs. The acquisition of these habits is provided by two kinds of training: "the professional training in basic principles that generally precedes entrance into organizational life, and the training through experience and planned job rotation that the organization itself can provide. Sometimes we supplement the latter with advanced management training in a university setting or a company training program" (ibid., p. 53).

[10]Ibid., p. 53.

shrink-wrap software industry, which prides itself on constant, if not at times, meaningless innovation, has steadfastly resisted the adoption of standardized software development procedures (see also Chapter 5).

Some Examples of Dubious Procedures

When I moved from Seattle to Southern California in mid-1997 I experienced some unusual difficulties with the local phone company. Prior to moving to Southern California, I had contacted the Seattle phone company to ensure that my Seattle phone numbers would be forwarded as of July 15 to my new numbers in California. After several calls to the Washington and California phone companies, I was assured July 7 that all had been taken care of. Imagine my surprise when on July 15 I decided to call my "old" number (to see if the call would be forwarded to my new number) only to hear a recording message informing me that my number had been disconnected!

What was particularly annoying about the incident was that every time I would call the phone companies (Washington and California), I had to listen to a prerecorded message assuring me that my call was important to the phone company. I was also told that "for purposes of quality control, my call may be monitored." When I would finally reach a human voice, I was told the same story: "Your number cannot be forwarded because you canceled your 'old' number." Each time, I would have to explain that the information entered on the computer was wrong and that I had not canceled the service. After approximately 10 minutes (during which time I could hear the operator typing some seemingly endless information) I was assured, as always, that the problem was now fixed and everything should be working within 24 hours.

This exercise in futility went on for approximately 9 or 10 days. Finally, on July 24 after *six separate requests*, my phone calls were at last properly routed.

What is revealing about this story is that the telephone operators were likely following a procedure, but apparently their procedure was not adequate to handle my perhaps unusual request (to have my old Washington numbers forwarded to my new California numbers). It was clear from the many dialogues that I had with numerous operators that most calls to the telephone company were calls to either turn on a service or cancel a service. Requests to have calls forwarded were apparently not as common, yet mine was certainly not the only request. Perhaps the operator should not have followed a procedure and really listened to my request. Perhaps the procedure did not clearly allow the operator to place a call forwarding request. For whatever reason, things did not work correctly.

Should You Ever Deviate from a Procedure?

The problem with procedures is not so much with the written instructions, but rather with the people who lack the wisdom to interpret the written instructions. Not long ago a football fan was arrested for selling two tickets to an undercover agent. His crime was not because he sold the two tickets to an undercover agent, but rather because he sold the tickets *for six cents over the legal profit margin*! A city ordinance in Jacksonville (where the crime was committed) specifies that people can sell (or resell) tickets for no more than 25 cents above face value. The tickets originally cost $34.72 each and the unfortunate man sold them for an even $35.00 each (3 cents each over the allowed profit margin!). The irony is that the man sued the city and eventually reached a settlement for $2500, which he donated to the football team.[11]

In some cases, following procedures can have deadly repercussions. An Associated Press report recently claimed that Swissair Flight 111 may have crashed on September 1998 off the coast of Nova Scotia because the copilot and pilot disagreed as to whether or not emergency procedures should be followed. The article claims that the copilot repeatedly suggested steps aimed at a quick landing, while the captain rejected or ignored those proposals "and became occupied with following a checklist of emergency procedures."[12]

The following anecdote was found in a 1997 *Los Angeles Times* article titled "Disney Not Amused by Jungle Skippers' Unscripted Jokes." The article described how a few of the captains on the Jungle Cruise ride dared to deviate from the official Disneyland script. The script, which apparently had been written a few years earlier, was, in the mind of some of the employees, dated and in need of some updating. To alleviate boredom and to make the ride more fun to customers, some employees began to slightly modify the script by adding their own jokes (sometimes even venturing into mild political jokes). When Disney management learned of the unauthorized deviations from the official script, they reacted promptly and fired the employees, some of whom had been working at Disneyland since 1988! One of the incredulous employees was quoted as saying, "People loved our jokes. I have never heard any complaints from the guests." Yes, but although the guests are the immediate customers, as far as Disney's management was concerned, the ultimate customer of all Disney employees is Disneyland and not Disney's customers. Apparently, the feudal Lords at Disneyland have no sense of humor and certainly do not believe in empowering their workers with any responsibility of judgment (perhaps a wise decision after all).[13]

Management consultant Peter Block (whose book *Stewardship* I highly recommend) would probably have disagreed with Disney's managers. Block explains that "The risk in relying too heavily on predictability is that basic purpose gets displaced. Strong emphasis on predicting outcomes drives our attention to things that can be measured. ... A heavy hand, however, leads people to give more attention to the measurement than to the service or product or outcome."[14] The demand for consistency and thus procedures at Disneyland

---

[11]*The Press-Enterprise*, December 29, 1998, p. D2.

[12]"Pilots Disagreed on How to Avoid '98 Swissair Disaster," *The Press-Enterprise*, January 22, 1999, p. A19.

[13]Shelby Grad, "Disney Not Amused by Jungle Skippers' Unscripted Jokes," *Los Angeles Times*, October 13, 1997, pp. A3, A25. Someone I know who works at Disneyland in Anaheim, California, explained that employees are actually allowed to deviate from the original script. Apparently these employees deviated too much and some tourists complained.

[14]Peter Block, *Stewardship*, Berrett-Koehler Publishers, San Francisco, 1993, p. 24.

clearly indicates that since rules cannot be broken, Disney's management is incapable of empowering its employees, for to do so would allow employees to deviate from scripted text—a risk Disney is apparently not willing to take.

Should All Processes Be Repeatable?

But should all processes be repeatable? In fact, are all processes perfectly repeatable? Some processes should *not be* repeatable and processes that are to be repeatable cannot always be absolutely repeatable because the raw material is not always consistent, which explains why operators often adjust process parameters such as temperature or rpms or feed rate and, in doing so, often must deviate from a standardized procedure.

Most people do not realize that a standardized procedure implies that certain assumptions are satisfied—assumptions that are rarely met in reality. The assumptions can be referred to as the *ceteris paribus* (or "all things being equal") condition often posited by scientists. An application of the *ceteris paribus* condition in the manufacturing world would be that steel 101-xx is *always* the same; an oven's temperature is always 450 degrees; all operators have equal skills, training, and experience; all equipment performs the same under any environmental condition; and all other conditions are perfectly repeatable or stable. If this is indeed true and if we further assume that no random event will disturb the process, then the output of the process would be repeatable (i.e., the same as before). This never happens, however.

So how can industry produce repeatable events? The fact is that many processes can tolerate variation. Process engineers, at least those with experience, allow some range in process specifications for key parameters. One must also recognize the continuous advancements in science, software development, and technology that have during the past 200 years led to the automation of processes and ultimate replacement of human operators with infallible and reliable motion-repeatable machines. These numerically controlled machines are usually intolerant of deviations-variations in specifications. Yet it is precisely these variations that have allowed some operators to maintain their jobs.

Prior to automation and even during the various stages of automation, processes were kept predictable (in control) thanks to human intervention. Human intervention consists of the millions of men and women throughout the world who have the ability, through their "craft-knowledge," to see variation or deviation from what should be and instantly adapt or modify their (micro) procedure(s) or numerically controlled machines to compensate for potential error(s). Examples of compensations (known as deviations) would include adjusting a temperature to slightly less or slightly more than called for in the procedure, tightening a bolt a little more or a little less depending on the feel, reinspecting a part that passed a test but does not feel or look right, and so on.

Pride in one's work is also an important factor. Most people would do whatever it takes to ensure that a process produces acceptable outputs. Describing the chaotic working conditions in Hotel X, George Orwell, who was to later achieve fame with his novels *Animal Farm* and *1984*, asserts that "What keeps a hotel going is the fact that the employees take a genuine pride in their work, beastly and silly though it is. If a man idles, the others soon find him out, and conspire against him to get him sacked."[15]

No matter how hard it tries, management will never be able to capture the infinite number of permutations that lead to the billions of decisions and deviations that are made each day by millions of workers throughout the world. In fact, the attempt to capture these decisions on paper has cost industries throughout the world billions of dollars. Yet this is precisely what science and technology attempt to do.

Thus, the easiest procedure to document is the procedure that is characterized by zero or near-zero variability. But because most processes have some variability, writing standardized procedures becomes difficult because one must make a series of assumptions that may not always recur. Stated differently, documented procedures generally work only if certain sets of predetermined conditions are satisfied each time the procedure needs to be repeated. For some processes the number of conditions (or degrees of freedom) are limited to two or three (i.e., easy-to-describe conditions) such as temperature, humidity, or type of raw material. But for most other processes, the number of degrees of freedom (i.e., the number of variables and combination of variables known as interactions that can affect a process) can be in the tens or hundreds of variables. Example of variables that could affect processes would include temperature, raw material, supplier, operator, and machine. Examples of interacting—or combination—of variables would include operator × machine, temperature × raw material, raw material × machine, or any other permutation of two- or three-way interactions of variables. For such processes, precise documentation would be a nightmare and may well be impossible. Still, thousands of such procedures have been written and will continue to be written. Yet variability can be significantly reduced with simple process modification. In one process, the variability in yarn thickness was found to be significantly reduced when a die that needed replacement was finally replaced. (*Note*: The die was not originally replaced because management was reluctant to invest the money to make it, and yet many hidden costs resulted from inspecting and reinspecting marginal ropes!)

Most people would argue that writing procedures is good because it defines and/or specifies how some tasks must be done, but are procedures always necessary under all conditions? As already explained, the written word was not always accepted as the best or most trusted way to describe a process. (In medieval times, the spoken word was preferred over the written word, which was perceived as being suspicious.)

Written procedures are certainly practical under certain conditions:

1. When a process is repetitive (i.e., predictable or stable) and
2. When a process is short such that it involves only a few (a dozen or so) steps.

---

[15]George Orwell, *Down and Out in Paris and London*, Harvest Books, New York, 1961, p. 75. See also Charles Hecksher, *White-Collar Blues*, Basic Books, New York, 1995.

Of course, for very complicated processes, procedures can be written as long as the complex process is broken down into a series (sometimes a very long series) of simple processes. Unfortunately, one of the consequences of simpler processes is monotonous jobs that require little or even no training and are also subject to high rates of turnover.

Technology has been very helpful when it comes to process documentation. In the chemical industry, for example, where processes are complicated, sophisticated and expensive programs control the process. The worker is often reduced to the passive role of watching screens and acting (once in a while) to change a blending ratio or increase temperature or some rate of flow. During the late 1990s, computer and communication technology has allowed processes to be described with pictures and are accessible by anyone with the appropriate computer password. Unfortunately, once again, companies seem reluctant to develop adequate systems. Thus, after having invested in all the necessary software to capture processes via pictures or film, some companies refuse to purchase large (17- to 19-inch) monitors. The result is that no one can decipher what appears on the small (15-inch) screen.

Many companies have failed miserably in their attempts to document (in writing) most processes. Although I have seen a few very good examples of process description, these are the exception rather than the rule. The overwhelming majority of the time, so-called process documents, say very little or say it so poorly that no one either understands or has the patience to read the volumes of documents. So why do companies still persist in wanting to document their processes the old-fashioned way (i.e., in writing)? In some cases, the blame can be placed on the ISO 9000 phenomenon. The standards require that just about everything be documented, but never really say why except to assume that documenting of processes is a good practice. But if that were true, how have companies managed to operate all of these years without religiously documenting their processes? The fact is that most companies have operated under the medieval premise that the word of mouth is better than the written word mostly because the written word often skips a few crucial steps and still cannot always be trusted.

What Is the Best Way to Document a Process?

How does one best learn to do something? What is the most efficient and effective way to demonstrate to someone how to do something? Would the trainee prefer to be handed a manual of procedures or instead be shown how to do a task? Which is better: a video showing how to assemble a part or a written or verbal description of how to assemble a part?

The amount of detail in a procedure should be correlated to the amount of training and/or skills required to perform the task and the level of risk (measured in terms of customer complaints, human-insurance cost, regulatory costs, or other financial costs) that could be incurred if something went wrong. Thus, the amount of detail found in operational procedures or emergency procedures written for nuclear plant operation is substantially higher than in a rope manufacturing plant where the level of complexity (and thus likelihood of forgetting important steps) and risk associated with errors is much less costly to the company, the customer, and society in general.

Can (Should?) Procedures Be Written for All Possible Scenarios?

The answer to the above question must be "no," because as the following story reveals it is not possible to envision and/or anticipate every scenario.[16]

A few years ago Steve Komarow wrote that during a military training exercise in Florida, about 100 soldiers set out in what should have been knee or thigh-high waters. But the Yellow River was rising and currents were strong. For some unknown reason, soldiers left their craft and began swimming in the strong current. Rescue efforts were hampered because of bad weather, and four Rangers died. Komarow commented that the ensuing investigative report noted that no standard operating procedure existed *for mass casualty evacuation from training.* Major General John Hendrix promised that a procedure would soon be written.

But what would such a procedure prevent in the future? And could they all be implemented in an emergency situation?

One must also recognize that, for certain work environments, the writing of procedures offer a considerable challenge. Long before achieving world fame, the novelist-journalist George Orwell experienced some very difficult times. In Paris, during the late 1920s, Orwell barely survived, sometimes not eating for several days, before landing a job as a *plongeur* (a dishwasher) in Hotel X. Orwell's vivid description of the hotel's hierarchical society as well as his commentary on the daily work routines is well worth the reading. "Anyone coming into the basement for the first time," Orwell recalls, "would have thought himself in a den of maniacs. It was only later, when I understood the working of a hotel, that I saw order in all this chaos."[17] A few pages later, Orwell returns to what he describes as the chaotic nature of a service quarters:

> *The thing that would astonish anyone coming for the first time into the service quarters of a hotel would be the fearful noise and disorder during the rush hours. It is something so different from the steady work in a shop or a factory that it looks at first sight like mere bad management. But it is really quite unavoidable, and for this reason. Hotel work is not particularly hard, but by its nature comes in rushes and cannot be economised. You cannot, for instance, grill a steak two hours before it is wanted; you have to wait to the last moment, by which time a mass of other work has accumulated, and then do it all together, in frantic haste.*[18]
>
> *You are, for example, making toast, when bang! down comes a service lift with an order for tea, rolls and three different kinds of jam, and simultaneously*

---

[16]The following story describing the military attitude regarding procedures is taken from Steve Komarow, "Nine Instructors to Be Disciplined in Ranger Deaths," *USA Today*, March 30, 1995, p. 2A.

[17]Orwell, *Down and Out*, p. 65.

[18]Ibid., p. 75.

bang! down comes another demanding scrambled eggs, coffee and grapefruit;
you run to the kitchen for the eggs and to the dining-room for the fruit, going
like lightning so as to be back before your toast burns, and having to remem-
ber about the tea and coffee, besides half a dozen other orders that are still
pending; and at the same time some waiter is following you and making
trouble about a lost bottle of soda-water, and you are arguing with him.[19]

One is curious to know how Frederick Taylor would have attempted to
scientifically analyze and optimize the service quarters of Hotel X. What type of
job specialization would he have created and would it have been more efficient?
What types of procedures could be written and dutifully followed for such a
work environment? Although one can assume that some of today's hotel/restau-
rants are probably run more efficiently than Hotel X, it is no wonder that only a
few hotels throughout the world have bothered to achieve ISO 9000 certification.
The task of documenting such a work environment must seem impossible to
master.[20]

Procedures versus "Show Me"

I once consulted with a small company (fewer than 15 employees) who strug-
gled for many months to develop procedures. Everyone seemed to agree that
procedures were needed, but nothing ever happened. The primary reason for the
delays and constant excuses was the process manager. It became evident after a
few meetings that the process manager did not want to document the half dozen
processes because to do so would diminish the control he exercised over his assistant. Indeed, he was the only one who really knew all of the processes and even
though he kept on assuring management that it was possible to document the
processes he never found the time to do it and always had more questions:

- How much needs to be written (details)? *Answer*: You don't have to write
  anything, we can use pictures or movies.
- How many pictures? *Answer*: I don't know. We will have to flowchart the
  process first and then decide how many pictures will be needed.

Once he finally agreed that taking pictures of the process would be easy he never
found the time to do it. Yet despite the unpleasant event, the experience taught
me a few valuable lessons and actually left me with even more questions than
answers. I began to wonder if one could truly transfer the knowledge of an expert
to a set of pictures and flow diagrams. The task is certainly theoretically possible
and would depend on the level of expertise, but even for what appears to be
simple processes with only four or five variables, the number of possible combinations of events that could go wrong increases rapidly.

---

[19]Ibid., p. 62.

[20]See James Lamprecht and Renato Ricci, *Padronizando O Sistema da Qualidade na Hotelaria
Mundial*, Qualitymark, Rio de Janeiro, 1997.

For the simple case of binary variables, that is, variables having only one of two settings possible, a process "controlled" by five variables could have as many as 32 combinations (or two to the fifth power). For variables having three possibilities (e.g., low, medium, and high), the combinations for five variables jumps to 243 possible scenarios. No matter how many or how few variables need to be controlled, if all possibilities are to be taken into account, the simplest of processes can very quickly become very long and expensive to document. Obviously, not all possibilities are documented—only the most likely ones.

Types of Procedures

The two basic types of procedures are (1) procedures that describe an assembly process that is repeatable and (2) procedures that (attempt) to describe a nonrepeatable process.

Procedures that describe an assembly process that is repeatable are the easiest procedures to write or capture in a blueprint, diagram, series of pictures, or video. Indeed, in such cases, the process relies on monotonous repetition of motion or tasks. However, even in such cases of supposed repeatability, deviations are often necessary. For example, when a supplier changes a formula to please another customer, the change(s) may have an unforeseen impact on your process (e.g., improved diodes, improved ink). Changes to improve a product may have costly consequences for others who have learned to adjust their product to a certain level of (reliable) quality.

Procedures that attempt to describe a nonrepeatable process have many IF branches in their flowcharts. Moreover the decision tree is not always as simple as "yes" or "no" because the answer to a particular question may well depend on other questions that need to be asked. In these cases, the person(s) in charge of the process needs training on how to adapt to various situations. Flexibility and the ability to acquire additional information are of essence.

What to Do?

The task is to first define the important conditions (boundary conditions) that are likely to impact the predictability of the process outcome. One effective tool to help define these important conditions is the flowchart.

- Identify the process boundary.
- Identify what comes into the process and what comes out.
- Create a flowchart of the process.
- Review the flowchart and begin asking "Why are we doing this here?" "Do we need an inspection point here?" If the answer is "yes because we often have rejects," then the next question should be "Do we have statistics to back up the claim?" What is meant by "often"? Can this be quantified?

Asking these questions will help you identify whether or not a problem is a real or perceived problem. If you have identified a real problem the next obvious questions are why do we have this problem and how long have we had this problem?

Also ask "Is this the most efficient way to do this task?" Find ways to simplify (see Chapter 16) the process and thus reduce or perhaps even eliminate the need for a procedure. Once a process has been documented you should always try to reduce its flowchart, not necessarily by reducing the boundary of the process, but rather by trying to simplify or eliminate or reduce the opportunities for errors. This can only be achieved by first charting the process and asking questions about it.

Other suggestions include these:

- The less educated the workforce, the more valuable videos or pictures. Yet the use of videos and or pictures to document procedures will also work for the most educated workforce.
- How much needs to be written? As little as is necessary to have the process work efficiently. Who decides process efficiency? You decide. There is no need to try to describe every contingency in your procedures because every procedure must be accompanied by some form of training that will complement the procedure.

Always remember the paradox of procedures:

- The more detailed a procedure, (a) the more likely someone will deviate from it, (b) the less likely someone will know all of the steps, and (c) the more likely it is to age and thus become outdated, unless the process is a stable, well-established process, which is unlikely to change.
- *Corollary*: The more detailed a procedure, (a) the more likely an auditor will find nonconformances.

Are Special Software Packages Needed to Document Processes?

During the late 1990s an increasing number of software packages have been marketed for the supposedly sole purpose of facilitating documentation efforts. Most people are anxious to purchase these software packages because they claim they do not have the time to develop their own system. This infatuation with software packages is puzzling because no matter how expensive or good (the two measures are not necessarily correlated) the software package is, it will not write procedures!

People who buy software packages fail to recognize that they will have to invest some time—and in some cases a lot of money—to learn how to use the software package. I have never understood the need to buy so-called ISO 9000 software packages because for most small to medium size companies, a very adequate software package already exists and is probably stored in your computer. Anyone who has Microsoft Office® or a similar package can use the many powerful features available in Microsoft Word®. A simple package such as Word® allows its users, among other things:

- To flowchart processes
- Use pictures or films to document processes
- Control documents and forms using passwords and a variety of other controlling features
- Insert voice annotations in any language to assist assemblers not fluent in English
- Use hypertext links to interconnect procedures

along with a host of other features designed to facilitate the documentation and operation of most quality assurance systems.

Naturally, to use such software packages efficiently, one must first thumb through the manual, read a few chapters, and learn to use its features. Unfortunately, because people do not apparently have the time to read, they prefer to buy another software package that they hope will require *no reading* whatsoever. Better yet, maybe the software package comes with a special Alt+Shift+Geni command that will instantly do everything.

Are Procedures Required for All Industries?

Ever since the publication of the ISO 9000 series of standards in 1987, it would seem that the manufacturing world cannot operate without procedures. Many people who begin implementing an ISO 9000-type quality assurance system readily admit that the development of procedures is an absolute necessity. Most would even confess that they have no idea how their company was able to deliver products without procedures. What has always puzzled me about these confessions is that prior to 1987—in the age of Before ISO 9000 (or BISO)—companies throughout the world were able to reliably produce quality products. Moreover, one would like to know how the industrial world was able to function prior to the ISO 9000 phenomenon! The fact is that, long before 1987, many companies had developed procedures (and many more, particularly in the software industry, had not). Unfortunately, in most cases, the events described by the procedure and the actual method of assembly did not always exactly match. Often, the "current practice" only resembled the written procedures.

Still, despite this obvious problem known as document control, good products were made and shipped to customers. Skeptics would immediately point out that although products were shipped to customers, many errors and hidden costs caused companies to be inefficient; profitability was not optimized and procedures would help the "bottom line." Optimists would not deny the charges but would also point out that more often than not, errors were avoided and, besides, there is no evidence that a fully "proceduralized" ISO 9000 registered firm is less likely to ship a bad part or product than an uncertified firm.

But all of the arguments for or against procedures fail to recognize one major point: Not all industries are created equal. Indeed, one cannot begin to debate the pros and cons of procedures without first attempting to categorize the various types of productions. In the early 1960s, Robert Blauner attempted to divide production technologies into four categories: craft, machine tending, assembly line, and continuous process.[21] Blauner's categorization is still valid today. Many processes are still based on craftsmanship and depend on the skill of the workers or master craftsman. Can the skill of all craftsmen be captured in a procedure and eventually codified or programmed for machines? That is precisely what industrialists have attempted to do, often successfully, from the dawn of industrialization. As soon as a skill could be proceduralized (the raison d'être of expert systems) and broken down into a set of subprocesses, the next logical step was to design a machine that could duplicate, as much as possible, the human skill. From the industrialists' point of view, the advantages of mechanization are obvious. However, although much mechanization has been implemented during the past 180 or so years, not all processes-activities can be proceduralized and eventually mechanized; some remain crafts.

In so-called "machine tending" activities, the skills of the machinist have been partly captured by the use of numerically controlled machines. The use of numerically controlled machines finds its origins in military requirements for very tight tolerances. As David Noble explains in his excellent *Forces of Production*:

> *The Air Force, in its development of high-performance fighter aircraft, was confronted with unprecedented machining requirements. The complex structural members of the new aircraft had to be fabricated to close dimensional tolerances and this extremely difficult and costly process seemed to defy traditional machining methods. The result was N/C technology, which also happened to meet military requirements for greater control over production (for quality control and "security" purposes) and manufacturing flexibility.*[22]

For machine tending activities, the use of procedures is therefore minimal because in most cases, the machinist generally punches in a program (an automated procedure) that has been approved and controlled by a programmer.

As far as the continuous process industry is concerned, key process procedures were programmed and thus automated in the dairy industry as early as 1925. By the late 1950s, the petroleum industry was already beginning to automate most of its processes. In the petrochemical and wood processing industries, for example, all key processes are controlled by multimillion dollar programs that display process flows on consoles monitored by operators. Surprisingly, these complex operating systems are often run by operators who are inadequately trained. On several occasions, the author has audited quality assurance systems where the operators readily admitted, much to management's displeasure, that they did not understand how the multimillion dollar computerized system they were supposed to operate functioned. When asked how they knew what to do, the answer often was "We've had some training and the rest we picked up by trial and error." Why management would want to purchase a million

---

[21]Robert Blauner, *Alienation and Freedom: The Factory Worker and His Industry*, University of Chicago Press, Chicago, 1964, p. 8.

[22]David Noble, *Forces of Production*, Oxford University Press, New York, 1986, p. 85.

dollar piece of equipment and then not properly train its operator has always puzzled me.

Shoshana Zuboff, who is also perplexed by this paradoxical behavior on the part of management, believes that the irrational behavior is partly due to "a profound underestimation (on the part of management) of skills demand associated with a technology that informates . . . it is almost as if managers preferred to deal with chronic dysfunction and suboptimal performance, rather than challenge the faith that held organization managers together in the covenant of imperative control."[23]

This leaves us with the assembly world. In most plants where some form of assembly is performed, a variety of procedures are usually found. The most common form of procedure/work-instruction is the assembly drawing, which depicts how various components and subcomponents are supposed to be assembled. Test and/or inspection procedures are also another common form of procedures found in assembly plants.

One of the fundamental assumptions erroneously made by most procedure writers is that the components or inputs feeding into a process are constant or standardized and thus predictably reliable over time. But how often is that true? And, more important, can it always be true? It all depends on how tight the tolerances, specifications, or product characteristics are set. Raw material extracted from the earth cannot always have identical characteristics. The more demanding tolerances and specifications are, the more difficult and costly it is to "pass product" as acceptable.

In the future, processes will have to be developed in such a way as to accept or, rather, tolerate variability. Processes and by extension products should be developed to absorb deviations from specifications and still perform as designed. Such products would dampen rather than amplify errors. Already, design engineers are experimenting with "fault or defect tolerant chips," that is, microchips "that can run anywhere and maybe even adapt on the fly."[24] As microelectronics are used more and more to control machines and thus processes, perhaps defect-tolerant processes are not so far away.

Should Procedures Be Written Like a Computer Program?

Before answering the question posed in the heading, one must first explain how computer programs are written. A computer program consists of a set of instructions—a code written in a particular language (C, C++, Fortran, Basic Plus, etc.)—designed to accomplish one or more tasks. A task may consist of a very simple proposition such as "Compute the square root of five." Nowadays, a programmer no longer has to write a program to compute a square root, he or she simply uses the SQRT (or some similar) command to calculate square roots. Still, years ago, someone had to write some code to compute square roots. Rather than reinvent

[23]Shoshana Zuboff, *In the Age of the Smart Machine*, Basic Books, New York, 1988, p. 254.

[24]Gary Taubes, "Evolving a Conscious Machine," *Discovery*, June 1998, p. 79.

the wheel every time, programmers long ago learned to rely on libraries of approved programs to compute certain tasks.

Suppose, however, that a programmer is asked to write a program that will compute payroll and print checks; this is a much more complicated task. Several input variables will have to be defined: employee name, code, salary, address, and so on. Numerous calculations will also have to be performed; for instance, the program will have to compute the state and federal tax rates based on the number of deductions, the Social Security tax and other payroll deductions, and the retirement fund rate, to name a few.

One characteristic of programs is that, once all the inputs have been defined, they expect to "receive" the correct information in the correct sequence in order to perform their tasks. If the information is not available, the software (i.e., program) may not work or, worse, compute erroneous information. A good programmer will insert various safety features to try to ensure that the software only produces an output after all inputs have been properly identified. Programmers must also try to anticipate unusual situations. For example, if the software user decides to do something unusual or unexpected, how will such an action impact the program? Of course, it is impossible to anticipate all possible actions.

Programs expect the information they need to be entered the same way every time, every day, year after year. Programs also expect information to be entered in a well-defined *sequence*; for example, name first, address second, and job code third. Any other sequence has the potential to "confuse" the program and produce erroneous or strange looking results. This dependency on proper sequence, repeatability-reliability with absolutely no variation, is usually not available when one describes work instructions or standard operating procedures. Work instructions depend on the ability of an operator, clerk, process controller, or manager to reach a decision when "things" (i.e., input variables) are not quite what they are supposed to be. People can also, depending on the nature of the procedure, vary the sequence or order of some steps without affecting the result-output. This is not true for software. Most computer programs, with the possible exception of artificial intelligence software, cannot "make decisions" based on inaccurate or insufficient information; they invariably break down and produce error messages or, worse yet, wrong information. That is one of the major differences between a computer software program and an individual performing a job that consists of many tasks and many more decisions. Programs do perform multiple tasks and are capable of reaching decisions using IF-THEN statements; however, for the most part, they reach dichotomous (yes–no) decisions (I don't want to get bogged-down with fuzzy set logic here). There is no room for "maybe," or "it's okay this time because we really need it," or "the specs are too tight anyway," or other types of fuzzy logic. People are often required to make such decisions on a daily basis. It is these types of decisions that are very hard to "program" in a rigid set of work instructions or standard operating procedures. These human decisions that are based on on-the-job experience (i.e., work knowledge) and an ability to learn and rationalize actions that can go beyond the scope of a procedure—sometimes with risky and costly consequences—are, to the best of my knowledge, unknown to computer programs.

Should we, or can we, force people to function as software, never deviating from a well-defined set of conditions? Or should we instead train people and give them the responsibility and authority that will allow them to confidently determine when they can operate outside the boundaries of a procedure or work instruction? My suggestion is that procedures or work instructions should be written, whenever possible, as guidelines. The employees responsible for performing a certain set of tasks as defined by their job description should, however, be either trained or have the proper combination of experience, education, and training to operate and function within the porous boundaries of the guidelines.

Summary and Conclusion to Parts III–V

Having reviewed some of the earlier attempts at regulating industries we are left with several obvious questions. In the case of 17th-century France, "Did the Age of Colbertism help French industry in the 18th or 19th century?" "Did the attempts at controlling and monitoring French manufacturers by forcing them in many cases to adopt official manufacturing procedures help promote French competitiveness abroad?" Finally, "Did Colbert's regulations help the overall quality of French products?" I would propose that the quality of French products was little affected by Colbert's noble efforts. Indeed, if one were to ask Americans to associate the word *quality* with French products, most would probably name the following items: wines, cheeses, perfume, champagne, food (in general), fashion, cosmetics, and probably art. Unfortunately for Colbert, none of these products (with the possible exception of wallpaper) attracted the attention of his edicts and auditors.[25]

Nonetheless, Colbert's desire to apply science and machinery to industry did eventually benefit the French since they were the first to create (in the early part of the 18th century) the best military engineering schools in the world, which were to later influence our own military academies.[26] It is difficult to demonstrate whether or not such early developments in military engineering helped the French mass produce, as early as 1785 (or before), muskets with interchangeable parts, but the fact remains that the technology that was to later influence Eli Whitney did originate in France.[27]

Still, although French entrepreneurs and manufacturers have achieved great technological successes leading to the production of high-quality products such as the high-speed TGV train and the French space program, most

---

[25]Wallpaper was developed in France in the late 17th century. The industry was probably controlled by Colbert's royal auditors. Colbert would no doubt be pleased to learn that a Charles X decorative screen was offered at New York Christie's for a mere $9000. Regulation does pay off, sooner or later.

[26]Artz explains that the word "engineer" is derived from the French "engins," or machines; hence the word *engenieur*. Frederick Artz, *The Development of Technical Education in France 1500–1850*, MIT Press, Cambridge, MA, 1966, pp. 47, 245.

[27]Lewis Mumford, *Technics and Civilization*, Harcourt, Brace and Company, New York, 1934, p. 93.

individuals do not perceive France as a nation of technocrats; that honor is generally bestowed on France's neighbor, Germany. Consequently and surprisingly, unbeknown to most Frenchmen, as far as overall perception of the quality of French products, France is still perceived as a nation of exquisite cuisine, excellent wines, elitist haute couture, and expensive perfumes. Colbert would probably be distressed to learn that the industries that he either had no knowledge of or had not even planned their control have successfully carved for themselves, and without much, if any, economic and "quality control" supervision, an impressive economic niche in the world markets based on the quality of their product. Ironically, France's recognition as a producer of quality products is, for the most part, due to the successes of the "regional aptitudes" alluded to by Boissonnade: cheese from the Brie region, wines from the Beaujolais and Rhone regions, champagnes from the Champagne region, perfumes from Grasse (a small town in the French southern Alps known for its many fragrant flowers), haute couture from Paris, and so on. French shipyards, iron (steel) making, textiles—the very industries protected by Colbert—have long ago lost their productivity leads and have been relegated to a secondary or even tertiary role in the world's economy. One must therefore conclude that Colbert's effort to enforce quality to protect consumers and revitalize French industry had met with partial success. Surely, government regulations were of no assistance in these cases.

The same could be said of major corporations that have had much experience working with the government and military specifications (however, the Boeing company was certainly favored by government contracts for the development of the 707). For example, when in the mid-1970s the Vertol division of Boeing began to design electric-powered light-rail vehicles for Massachusetts Bay Transportation Authority (MBTA), it quickly found out that the knowledge acquired designing helicopters did not easily transfer to public transport. After repeated operational failures, many directly linked to inadequate designs, "the MBTA and Boeing-Vertol arranged a legal settlement under which Boeing-Vertol agreed to pay MBTA $40 million as final settlement. . . ."[28] Obviously, the experience acquired from military projects did not prove helpful for the Vertol division of Boeing. It is interesting to observe that in 1994, Raytheon, maker of Patriot missiles, had decided to enter the rail-car business. Interestingly enough, Raytheon was seeking contracts to supply trolley cars to Boston and high-speed trains for Amtrak's Northeast corridor. Before history is to repeat itself, someone in Boston should talk to the Massachusetts Bay Transportation Authority (MBTA).[29]

To the credit of some military leaders, the limitations of traditional mass production had already been recognized some years ago by certain individuals within the military. Recognizing that the homogeneity of Second Wave society will be replaced by the "heterogeneity of Third Wave civilization," Alvin and Heidi Toffler wrote that "De-massified production—short runs of highly customized

---

[28]Seymour Maelman, *Profit without Production*, Alfred A. Knopf, New York, 1983, p. 256.

[29]Digest section, *USA Today*, September 28, 1994, p. 3B.

products—is the cutting edge of manufacture."[30] Short customized runs may well be the best approach for some industries, but research by Robert S. Kaplan and Robin Cooper has demonstrated that in many cases companies produce too many products. Using activity-based costing (ABC), researchers at Harvard discovered the "whale-shaped" profitability curve. The curve demonstrates that "The most profitable 20% of products can generate about 300% of profits. The remaining 80% of products either are break even or loss items, and collectively lose 200% of profits, leaving the division with its 100% of profits. This curve happens so frequently in ABC analysis that it has been given a name, the 'whale curve.' "[31] Profitability is not addressed in the ISO 9001-2000 standard and, yet, as any businessperson knows, profitability is the very raison d'être of any business.

If one is to explain why certain companies and their products have achieved worldwide recognition for their quality, factors other than antiquated standards must be considered. To use an old cliché, a different theory or paradigm must be found. The current paradigms offered by the quality profession (unfortunately, there are many; perhaps too many) are too myopic in scope for they have failed to recognize that although quality in all its vagueness and occasional triteness is certainly an important variable to consider, it no longer is (if it ever was) the only or even the predominant concept to enforce. Quality without economic viability or survival is a useless concept; this has long been understood by most practical businesspeople, but, as we shall see, by very few quality experts. Companies have established regional or global dominance not solely because their management attended ISO 9000 seminars, leadership seminars, or TQM seminars, but rather because of economic, financial, and other intangible criteria that go beyond the techniques proposed by quality practitioners.

---

[30]The Tofflers also quote Don Morelli of the DoD who had recognized more than a decade ago that the military was also moving toward "the de-massification of DE-struction in parallel with the de-massification of PRO-duction." Alvin and Heidi Toffler, *War and Anti-War*, Little, Brown and Company, New York, 1993, pp. 22–23, 72. The press has covered extensively how the defense industry is restructuring itself; see, for example, Phil Hampton, "$10 Billion Deal to Join Defense Firms," *USA Today*, August 30, 1994, pp. B1–B2, and Janet L. Fix and Eric D. Randall, "Defense Giants to Merge," *USA Today*, August 31, 1994, pp. B1–B2.

[31]Robert S. Kaplan and Robin Cooper, *Cost and Effect*, Harvard Business School, Boston, 1997, p. 162.

Part VI

Conclusion

The following classic example of "friendly tyranny," which sets quality professionals in opposition to other members of an organization, summarizes not only the daily confrontations faced by countless people throughout the world but also exemplifies the essence of what has been described in previous chapters.

I was training four employees on how to conduct internal audits. The four employees represented engineering, quality, manufacturing, and purchasing. During the course of the audit the five of us (myself plus the four trainees) arrived at an assembly work station. Everything was going well and we were about to conclude our audit when, in an attempt to explain the design process, the engineer stated that a couple of electrical engineers did not modify (and thus formally release) engineering drawings until after they were sure that they had achieved the right configuration.

On learning of this "confession" the quality manager proceeded to cross-examine the engineer and attempted to convince him that such procedures were unacceptable because it clearly demonstrated that the engineers in question had no control of the design process. "They should formally use our engineering change notice procedure every time they make a change to the part," the quality manager explained. "But that is not practical," the engineer retorted. "They may make four or five changes a day before they get it right. And besides, you've got to remember that by the time they get the part we often are slightly behind schedule and they have to work very hard to catch up and make sure that we satisfy our delivery date."

This did not impress the quality manager who insisted that even if they made 10 changes a day the electrical engineer had to formally document all of their changes, every day. The conversation went on for about 15 minutes. The quality manager maintained that the electrical engineers had no control of their process, and the engineer maintained that they had absolute control because the part never left the engineer's hand until a solution to the design problem was resolved at which time a drawing was formally released to manufacturing. "Why would manufacturing want to know about all of our trials and errors?," was what the engineer seemed to be saying.

Realizing that the company rarely made the same identical product twice, I suggested that perhaps we were dealing with a case of prototyping. In other words the electrical engineers had decided to use this particular method of documenting only their last (working) change precisely because they knew that the documentation of all experimental trials was too time consuming and not worth their effort. The engineer liked my suggestion, but the quality manager was not convinced. "Information was lost," he proposed, which was true, but how much information would one need to monitor and who would look at all the changes?

As I listened to the debate which was now approaching 30 minutes long, I began to side with the engineer; however, noting that no one had asked for my opinion, I decided to keep quiet. The engineer finally asked the quality manager, "Please explain how your method would add value to our process. I would like to help but I don't know how." The quality manager remained silent and never answered the engineer. His point was that "This is not right," but he could not really explain and thus convince the engineer why his way was better. Certainly his way was more time consuming and the engineer wanted to know how all this additional documentation was going to benefit anyone, except perhaps quality.

It is likely that different people will react differently to the above story. Quality professionals will side with the quality manager and probably cite a long list of reasons as to why their method is better (few if any of those reasons will likely invoke an economic or financial argument). Engineers will probably agree with the engineers and cite all sorts of efficiency statistics to prove their point. The answer would require a more detailed analysis that, unfortunately, no company is likely to be willing to sponsor.

The above scenario is not unique to any industry and is mild when compared to other industries. A consultant friend who specializes in the increasingly prescriptive automotive standard known as QS 9000 told me that suppliers are beginning to rebel against the unreasonable demands imposed by their insatiable customers. The difference of opinion between quality and engineering or quality and manufacturing as to what is practical, feasible, efficient, valuable, or justifiable has existed for a very long time. Still, the above story does summarize the clash between two realities: reality as perceived by quality professionals and the separate reality perceived by nonquality professionals. What is particularly amusing about the above story is that the quality manager and the engineer both had several copies of Scott Adams' *Dilbert* cartoons (sometimes the same cartoon) pinned on their office walls or inserted in some of their manuals! Most of these cartoons related to the ISO 9000 standards. Ironically, neither the engineer nor the quality manager realized that their confrontation was classic Dilbert material.

CHAPTER 16

By Way of Conclusion: Dos and Don'ts

Challenges for the Quality Professional in the 21st Century

As we enter the 21st century, the quality profession and businesses in general will be faced with new challenges. There is little doubt that quality professionals will have to reinvent themselves or face eventual elimination as a profession. No longer will they be able to pontificate about quality for the sake of quality. For quality professionals, one of the bitter ironies of their success at introducing fundamental quality principles to a broad range of industries is that their success will likely lead to the eventual eradication of their profession. Nonetheless, because managers will continue to be needed for a few more years, the quality profession, as we know it today, is not likely to disappear overnight.

What is likely to happen or perhaps should happen is that during the next few years the responsibilities currently associated with qualiticiens or quality professionals will eventually be absorbed by other professions and will become part of the daily activities of every "worker." When this transition is complete, the quality profession, which is known today as a separate entity, will no longer exist as such but will continue to exist in spirit spread across numerous activities and/or responsibilities.

On the research front, the quality profession and businesses in general need to rethink the paradigm of minimum variability. Instead of focusing solely on minimizing variability (of product, for example), researchers in the field of quality should focus on how to exploit maximum allowable variation (noticing that parts that do not need to be replaced need not be interchangeable; Radford suggested something similar more than 70 years ago).

Need to Integrate Many Methods

The next challenge for the quality profession and business in general will be to integrate various methodologies (ISO 9000, six sigma, activity base accounting, etc.). Although there is already a general, if somewhat vague, awareness about achieving methodological integration, to my knowledge, no serious effort or research has begun in this field.

The need for integration is evident when one notices the convergence between methodologies. For example, the ISO 9001-2000 standard now includes more requirements for data collection, process monitoring, data analysis, and continuous improvement. These new requirements brings the standard closer to the popular six-sigma methodology rediscovered in 1998. Similarly, the need to establish measurable objectives to assess the effectiveness of an organization's quality policy and the new requirements placed on assessing customer satisfaction and dissatisfaction bring the ISO 9001-2000 standard closer to the Malcolm Baldrige award.

Yet, built within these standards are the conservative seeds of stability and status quo. There is nothing new in the ISO 9001-2000 standard that has not already been said decades ago. This is to be expected of any standard. Standards do not promote innovation; they merely try to ensure (some would say impose) that the status quo is adopted by more and more people. One of the possible ironies of the standardization and certification processes is that those left outside the standardization-certification process may actually outperform companies that have achieved certification for the sake of quality! This could happen not because standards promote bad practices (for they do not), but, rather, because companies left outside the brotherhood of (ISO) certified companies may be more adept at responding to change. A case in point might be the software industry and the generally accepted concept of "good enough" quality.

The philosophy of "good enough" quality has long been practiced and defended by many software developers. One of the most eloquent defenders of the "good enough" philosophy has been James Bach. Bach's "good enough" thesis is based on some fundamental assumptions about quality, generally software quality, that I propose apply equally well to other industries. In a style reminiscent of Pirsig (see Chapter 5), Bach proposes that quality is not mere substance but rather, an optical illusion, "a complex abstraction that emerges partly from the observed, partly from the observer, and partly from the process of observation itself."[1] Building on that premise, Bach next proposes that clients (whether they are internal or external), never really know the quality of a (software) product. To prove his point, Bach observes that in many cases of software development, the customer "viewpoint" is not necessarily known or is perhaps even unachievable. Moreover, Bach asks, "How do we incorporate their (customers') values into a product they haven't seen yet?" This same dilemma is faced by many other industries who are market driven for their product innovation.

What Bach suggests is that clients form a perception of software quality. To further complicate matters, one must recognize that this perception of quality in turn depends on customer values, their skill level, and past experience. The challenge for the software developer is to develop a product and hope that the customer (quality) perceptions will match the developer's perception.

Given the ill-defined nature of the challenge, software developers are faced with the difficult task of trying to identify the appropriate quality factors, adequately measuring these factors, and determining how to control them. To accom-

---

[1] James Bach's article can be found at http://www.stlabs.com.

plish these tasks, Bach proposes a utilitarian view of quality that he summarizes as follows: "The quality of something should be considered good enough when the potential positive consequences of creating or employing it acceptably outweigh the potential negatives in the judgment of key stakeholders."[2] Bach goes on to propose five key process ideas (KPIs) to create good enough software: (1) a utilitarian strategy, which includes risk management, decision, game and control theories, and principles of fuzzy logic; (2) an evolutionary strategy, that is, the ability to respond to changing requirements; (3) heroic teams, which are needed to solve ambiguous problems; (4) a dynamic infrastructure, which will create the ability to respond to the needs of a project; and (5) dynamic processes that can change with each changing situation. Bach's recommendations would apply to most other industries.[3]

Challenges for Companies in the 21st Century

The paradox of ISO 9001-2000 or any quality standard focusing on documentation is that it tends to promote bureaucracy and it does not really guarantee success in the marketplace—software designed to help companies cope with the documentation and document control process does not ease but, instead, enhances the bureaucratic burden. One unfortunate side effect of too much "proceduralizing" is that it can drive away creative, innovative people, the very people who can help a company develop new products and stay one step ahead of the competition.

One of the legacies of globalization promoted by the World Trade Organization and others is the continued wave of mega-mergers. The growth of mergers and acquisitions clearly indicates that major corporations have come to the conclusion some years ago that, in order to survive, they must get bigger and thus acquire larger and larger world market shares. As multinationals transform themselves into gigantic mega-firms, thus raising entry barriers and simultaneously reducing competition, what will be the impact on the overall quality of the products they produce for society? Will these firms eventually become immune to the fundamental principles of quality practiced during the past 50 years? Until the economic world is reduced to a handful of gargantuan firms that will own and distribute just about everything, one can assume that mega-firms will continue to produce quality products. In fact, it is likely that the bigger the firm, the greater the chances it will be able to rely on technological improvements to reliably produce quality products. Their ability to effectively handle customer complaints will probably still remain an elusive goal.

It is difficult to predict what will happen, but by observing certain industries, particularly the software industry still dominated by Microsoft, one can venture a guess as to what will happen to product quality in the future. As firms become larger and control a larger share of the market they could more easily impose their level of "good enough" quality on customers. Microsoft, for example, has

---

[2]Ibid.

[3]Ibid. Bach's article provides further details.

over the years produced many "good enough" software packages and has even managed to convince customers (at least for now) that releasing new versions of software is more important than releasing bug-free software. This is not necessarily a bad option for consumers unless, of course, overall quality of product and service declines to abysmal levels, something that is a common practice in the former Communist bloc countries, for example. One would hope that the emerging e-business market, which already grosses billions of dollars, will not follow that path. Based on current practices, the overwhelming majority of Internet sites still have a very long way to go when it comes to providing minimum customer service. One could even suggest that the very essence of Internet business is to be customer *un*friendly.

Final Thoughts on Don'ts

Some years ago Dr. Juran observed that "the quality industry created much history but few historians."[4] Having always been interested in history, particularly economic history, I could not agree more with Dr. Juran. What I have tried to do in the previous 15 chapters is demonstrate that historical analysis is essential in understanding not only the evolution of ideas but also why they are periodically recycled and peddled to a welcoming audience supposedly for the sake of quality.

In the same article Juran also wrote of the need to have "the moral courage to be different when it would be so easy to conform to long-standing tradition."[5] That part was easy for me. I have never been able to relate or associate with quality professionals whose vocabulary consists of the latest platitudes they have either read in a popular book or heard at the latest American Society for Quality (ASQ) conference. The world of quality professionals is full of these annoying trite statements that are repeated *ad nauseum* by consultants, auditors, and quality engineers: "ISO 9000 is a minimum standard"; "It will help level the playing field"; "TQM is of course better"; "You can't have quality without procedures"; "If you don't have records then it (event) did not happen"; "What would happen if this employee were hit by a bus?"; "Can you recite your company's quality policy?"; "Have you calibrated your measuring tape?"; and so on.

What is more disturbing is that in some cases, people will agree to do the most idiotic things just to please a demanding auditor. Usually they do these things out of fear or simply to get the auditor off their backs and get back to the business of earning a living. The many incidences of irrationality that I witnessed during the 1990s ensured that I could not possibly be associated with the triteness of the status quo. So what have we learned in this book?

At the risk of falling into the trap of triteness I will state but two simple recommendations:

---

[4]Dr. Juran wrote: "This future medalist will, incidentally, have his work cut out for him, since industry creates much history but few historians" ("Pioneering in Quality Control," *Industrial Quality*, September 1962, p. 14).

[5]Ibid.

1. *Develop better ways to measure and optimize your cost systems to help you reduce and eliminate waste.* This advice, which has been suggested for decades, may be easier said than done, but one way to start optimizing costs and improve efficiency is to stop buying "how to" books on management and quality that promise quick solutions and start reading books that offer a sounder methodology that requires some effort and thought to implement. The quality literature has published a few books on the subject of quality costs, but it's surprisingly few. Of the few books that have been published, Juran's *Quality Control Handbook* remains, in my opinion, the best source.[6]

During the 1990s, many scholars expounded on the subject of quality costs. Out of their research a promising field of investigation has emerged: activity-based costing (ABC). The literature on activity-based costing is vast and growing rapidly. Several books have been written on the subject of activity-based costing. One of the best is Robert Kaplan and Robin Cooper's *Cost & Effect.*[7]

If you are serious about reducing waste and do not know where to start, may I suggest that you start looking at your nonconformance reports (NCRs) and corrective actions (CAs) (assuming of course you keep such records). After perusing the last 3 to 4 weeks of NCRs and CAs, you may be surprised at the magnitude of what used to be called the "hidden factory." As you look at these reports, I would also suggest that rather than trying to justify each and every one of them you try to find ways to *eliminate* the problem(s). Although it is true that some of these "problems" will have unique scenarios, my experience suggests that in many cases broad patterns of nonconformities and corrective actions can often be detected thus indicating an endemic problem begging to be resolved.

Optimizing also means optimizing the multitude of software systems your company may currently be using. Over the years, I have often been surprised by the lack of structure of many companies' information "management" systems. Too many companies have a plethora of software systems in place and the "management" of these systems leaves a lot to be desired mostly because software is purchased *ad hoc* instead of planned for integration. In many cases it is difficult or impossible to transfer data from one department to the other. Incredibly enough, data cannot be transferred because in many cases (especially small companies), the computers are not connected to a local-area network. When a system does exist, the employees may not feel comfortable with the system and do not always know how to use all of its features.

2. *Whenever possible, recognize symptoms of complexity and try to simplify your processes.* This piece of advice should be easy to follow, but unfortunately the age of ISO 9000 has led many companies to develop—often with the "assistance" of specialized ISO 9000 software—so many cumbersome procedures that simplicity is almost perceived to be an obscene concept. To simplify a process,

---

[6] J. M. Juran and Frank M. Gyrna (Eds.), *Juran's Quality Control Handbook* (various editions), McGraw-Hill, New York, 1988.

[7] Robert S. Kaplan and Robin Cooper, *Cost & Effect*, Harvard University School Press, Boston, 1998. For a history of management accounting and its irrelevance to today's corporate needs, see H. Thomas Johnson and Robert S. Kaplan, *Relevance Lost: The Rise and Fall of Management Accounting*, Harvard Business School Press, Boston, 1987.

one must first recognize its complexity. Most people do not perceive the processes they work with as being complex (or even irrational) because they have worked with these processes for some time and have gotten used to them or because certain processes acquired their complexity slowly over time. To help people identify some of the symptoms of complexity, I have prepared the following list:

Symptom 1: Repeated occurrence of errors with no attempt (no time) to solve problem.

Symptom 2: Procedures that are complicated and encourage errors.

Symptom 3: A bureaucratic quality assurance system that does not match current resources (which may have been downsized over the years).

Symptom 4: Belief that more is better: more documents, more signatures, more procedures, more detail, more inspection, more control, more forms or longer forms, and so on.

Do not develop mechanistic procedures as if they were commandments that are to be enforced by authoritarian internal auditors. Determining the correct amount of detail in any procedure is a difficult task. In attempting to define the proper amount of information to be contained within a procedure always remember Herbert Simon's concept of man as a satisfier agent (i.e., the "good enough" principle).

Also recognize that the problem of proper or adequate documentation can be viewed as a "wicked problem." The concept of a wicked problem originated with Horst Ritter in the early 1970s when he was working on developing a theory of design. A design problem is said to be wicked when "it is ill-formulated, where the information is confusing, where there are many clients and decision-makers with conflicting values, and where the ramifications in the whole system are thoroughly confusing."[8] Often, stating the problem is the problem. The design of procedures can easily deteriorate into a wicked problems when the "designer" (i.e., procedure writer) first attempts to determine whether or not a procedure needs to be written in the first place. Having decided that a procedure would be written, the designer is next faced with the mistaken belief that the content of the procedure must simultaneously satisfy (internal or external third-party) auditors as well as the user of the procedure. When developing procedures, do not worry about the auditors; instead, worry about whether or not they are needed and useful and able to help you improve a process.

Now, having identified complexity, how can one simplify?

How to Simplify

Let's take a look at some steps one can take to help simplify a complex process:

- *Minimize approvals.* Keep the number of approvals (signatures) required at a minimum. More than one approval reduces control by diluting accountability.

---

[8]Horst Ritter quoted in Paul Coyle, "The Madness of My Method," http:www.uclan.ac.uk/facs/destech/viscomm/pess.htm.

- *Reduce system/process complexity.* This can be achieved by recognizing excessive training as a symptom of complexity.
- *Develop processes and procedures that will require minimum training.* If training cannot be avoided, it should not exceed more than a few hours.

Negative side effects of these steps include marginalization or alienation of workers, which could affect the quality of work life. Process simplification, however, need not necessarily mean a reduction in responsibility. As jobs are simplified perhaps people need to be given more responsibilities. This is what has happened with downsizing; however, without the simplification part!

Jobs that are simple and repetitive in nature tend to be very low paying. They also tend to be boring jobs. This in turn leads to high absenteeism and high turnover (300 to 400 percent per year). In the assembly world, many jobs have been replaced by high-performance machines. The operator is needed merely to verify that the machine is functioning properly.

Additional steps to reduce complexity include these:

- *Reduce the number of forms and simplify forms whenever possible.* (Suggestion: Look for forms that have been only partly completed and ask yourself if the information is required or can it be eliminated.)
- When a form is redesigned, test it for a few weeks to see if more information can be removed or if more information needs to be added.
- Do not try to anticipate all possible options (this makes for complicated forms). Instead, include a "Comment" or "Option" area.
- Use the back of the form to include additional information and/or instructions on how to fill it out.
- Train people on the use of forms.
- Plan your system, develop it, try it, and redesign it (simplify it).
- Always ask yourself: "What is the value added to this step by this process?"
- Use the rule of the Five Whys? Ask "Why are we doing this?" and keep asking. For example: "Why do we need a purchasing order?"
  "Because the purchasing department requires us to do so."
  "Why does the purchasing department requires us to use a form?"
  "Because . . . ."
  "Why . . . ?"
  Usually, by the third or fourth "why" you will find the root cause. You must then ask: "Why is that root cause present? Can we change it or eliminate it?"
- Whenever possible, use pictures or movies instead of written documents to document processes.
- Use software packages (Adobe Premier, for example) to document processes (via pictures or movies) and train workers. One possible negative side effect is that software packages do not always simplify life.
- Develop precise and short job descriptions that specify responsibilities and/or authority.
- Develop meaningful performance measures but limit them to two or three key variables per department (i.e., internal customers).

- If performance variables are not assessed during management reviews then they are probably not important and should be discarded.
- Use the corrective action process to uncover repetitive problems. If the same or similar errors are repeated, use the rule of the "Five Whys" to try to identify the fundamental cause. Next, see if a process simplification could eliminate the problem. This may require redesigning a process (assembly, training, testing) or combination of processes or redesigning the product or system.
- Processes may have to be combined to reduce-eliminate errors.
- Try to adopt the fundamental principle of "less is more."

Having offered some suggestions as to what you should do, I will now list a few things you should avoid doing.

ISO 9000 Software: No Panacea

Do not believe in the quick fix or software packages that promise you efficiency. Software cannot help you become more efficient if you are not efficient or do not know how to organize your requirements. Software packages can help you to become better organized but that is only true if you are already inclined to be organized. As much as some people would like to believe it, software cannot perform magic. Don't think that software packages will help you achieve ISO 9000 certification. That is still impossible.

On Quality-Speak

Over the years, quality professionals have developed their own jargon. This quality-speak reminds me of George Orwell's Newspeak, which he mentions in his famous novel *1984*. Avoid prophets who speak in quality-speak. Whenever in the presence of quality prophets, try not to listen; if you must listen, consider everything with some skepticism. This is not to say that all quality prophets do not occasionally have something worthwhile to say. Some do, but always wonder if there might not be an ulterior motive.

Don't believe everything you hear about new paradigms or methodologies; some of it is either of dubious value to your business or is recycled and repackaged old ideas (old ideas are not necessarily bad, however). Always consider the source of the information. Remember that the vast industry of quality is in the business of promoting new fads because part of its income is generated from seminars, book publishing, and conferences. No one fad will help you solve your problems; instead, a series of good, commonsense practices, consistent effort, an ability to listen, and a willingness to adapt and react to the needs of the market are the well-tested practices that are likely to help you, at least for a while.

Recognize what should be obvious: that "quality" is but one of the many constraints involved in the design of a product. In the case of some industries, such as the software industry, quality may not even be one of the most important constraints. Researchers in marketing recently confirmed what has been known by many: "When making a decision between brands customers may rationally choose the familiar brand with worse expected quality, even if all non-quality attributes are equal between the two items."[9]

Remember that nothing lasts forever; change is the only constant. No matter how stable your processes, they will drift or become outdated with time. Therefore, you will periodically need to adjust and readjust your processes. Even if your processes do not exhibit characteristics of instability, your competitors may come up with innovations that will force you to react and change.

Teamwork: Another View

Many companies firmly believe in the teamwork process to resolve any and all problems and, yet, as some sociologists and anthropologists have discovered, teams are far from being a perfect or efficient mechanism to resolve problems. In some cultures, the concept of a team can have a negative connotation. For example, the anthropologist Charles Darrah noted that to Vietnamese, working in a U.S. assembly plant, the concept of a team was difficult to accept because it reminded them of Communist work teams—something they thought they had left behind when they fled Vietnam.[10]

Workers joining teams—defined by the sociologist Richard Sennett as a group of people assembled to perform a specific, immediate, task rather than to dwell together as in a village—must have an instant ability to work well with a shifting cast of characters. "That means," Sennett observes, that "the social skills people bring to work are *portable*: you listen well and help others, as you move from team to team, as the personnel of teams shifts—as though moving from window to window on a computer screen."[11]

Some sociologists have found teamwork to be a kind of deep acting "because it obliges individuals to manipulate their appearances and behavior with others . . . the art of feigning in teamwork is to behave as though one were addressing only other employees, as though the boss weren't really watching."[12] The successful players rarely behave the same way off screen, when the boss is not watching. But perhaps the worse condemnation of teams was offered some years ago by Lewis Coser who argued that "people are bound together more by verbal conflict than by verbal agreement, at least immediate agreement . . . there is no community until differences are acknowledged within it. Teamwork, for instance, does not acknowledge differences in privilege or power, and so is a weak form of community; all the members of the work team are supposed to share a common motivation, and precisely that assumption weakens real

---

[9]*Quality Progress*, October 1999, p. 21.

[10]Charles Darrah quoted in Richard Sennett, *The Corrosion of Character*, W. W. Norton & Company, New York, 1998, p. 111. William Wolman and Anne Colamosca note that "those that did not see themselves as team workers were denounced—much as in China during the Cultural Revolution—as reactionaries who had not honed their skills for the revolution" (*The Judas Economy: The Triumph of Capital and the Betrayal of Work*, Addison Wesley, Reading, MA, 1997, p. 61).

[11]Sennett, *The Corrosion of Character*, p. 110.

[12]Ibid., p. 112.

communication. Strong bonding between people means engaging over time their differences."[13]

To Dave Arnott, author of *Corporate Cults*, the process of team building often consists of "enforced bonding," whereby executives or employees are strongly encouraged to spend a few days at a retreat to learn about meditation, the benefits of a vegetarian diet, and, most importantly, workplace "togetherness." To Arnott, this kind of personal invasion disguised as team building is clearly cultic.[14]

Having had to participate in various teams over the years, I must agree that I rarely found them to be effective. My experience with teams has been that although small (three to five individuals) teams can on occasion be effective, large teams tend to complicate, overproceduralize, and overanalyze the simplest of problems, problems that could have been resolved in a fraction of the time by one or two individuals. Let us hope your company is not too infatuated with teams.

Smaller Would Be Better

In 1957, long before the breakdown of the Soviet Union, Yugoslavia, or Czechoslovakia, the late Austrian economist Leopold Kohr (1906–1994) published a fascinating book titled *The Breakdown of Nations*. In his introduction, Kohr writes that the cause behind all forms of social misery is bigness. "Wherever something is wrong," Kohr writes, "something is too big."[15] To restore balance, Kohr suggests the breakdown of overgrown superpowers and a return to "small and easily manageable states such as characterized in earlier ages."[16]

Applying the same rationale to business units, Kohr observes that some economists, "are rediscovering the value of the small-unit principle and suggest that a multicellular arrangement with as many independent entrepreneurs as economically supportable would be more wholesome, productive, efficient, and profitable than a world composed of giant concerns spilling across the surface of the globe, unimpeded by limiting boundaries."[17]

John Hagel and Marc Singer recently reached a similar conclusion.[18] Singer and Hagel observe that when one looks beneath the surface of most companies, one finds three kinds of "core practices" *bundled* under one organization: a customer relationship business, a product innovation business, and an infrastructure

---

[13]Coser quoted in Sennett, *The Corrosion of Character*, p. 143.

[14]Dave Arnott, *Corporate Cults: The Insidious Lure of the All-Consuming Organization*, Amacom, New York, 1999, p. 81.

[15]Leopold Kohr, *The Breakdown of Nations*, E. P. Dutton, New York, 1957, 1978, p. xviii.

[16]Ibid., p. xix. When writing about bigness, Kohr was not referring to an absolute universal size beyond which all units would have to be broken down. To Kohr bigness was a relative concept. His concept of national bigness did not only apply to the Soviet Union and the United States but also to smaller nations such as France, Italy, and even Switzerland. Although not using the term, Kohr was referring to ethnonationalism.

[17]Ibid., p. 164.

[18]John Hagel and Marc Singer, "Unbundling the Corporation," *Harvard Business Review*, 77, March–April 1999, pp. 133–143.

business. In an effort to optimize overall performance, large corporations have, during the past decade, spent a lot of time and effort on reengineering or redesigning their (bundled) core processes. But, as Singer and Hagel observe, "managers have found that there are limits to such gains. . . . Bundling (core processes) into a single corporation inevitably forces management to compromise the performance of each process in ways that no amount of reengineering can overcome."[19] The difficulty in achieving efficiency, the authors explain, is partly due to the fact that the three types of core practices have different objectives. To see what the organization of the 21st century is likely to look like, the authors present several case studies of Internet companies. Their conclusion is that "While traditional companies strive to keep their core processes bundled together, highly specialized competitors are emerging that can optimize the particular activity they perform. Because they don't have to make compromises, these specialists have enormous advantages over integrated companies."[20]

Unfortunately, based on the seemingly unending wave of national and international multi-billion dollar mergers that have occurred and continue to occur to this day, one would have to conclude that corporate executives are either unaware of Kohr's thesis or simply do not believe in it. The many international committees that are forever producing more and more international guidelines and standards as well as the ever expanding industry of standards such as those produced by the automotive industry also sadly demonstrate that the cycle of growth has not yet ended. The belief that more and bigger are better for the current "global order" will no doubt reign for a while longer.

Meanwhile we must patiently await the age when gargantuan global multinational companies will be broken down, preferably by market forces rather than government edict, into smaller more rational business units. When that day comes, suppliers will be able to throw away their international certifications or simply not renew them and go back to what they have always done: Produce quality products without the bureaucratic nonsense demanded of them by unreasonable and all too powerful purchasing managers working for all too powerful organizations.

---

[19]Ibid., p. 135.

[20]Ibid., p. 136. The authors coin the term "infomediaries" for these specialized competitors who collect customer information and customize products to their needs. See also John Hagel and Marc Singer, *Net Worth: Shaping Markets When Customers Make the Rules*, Harvard Business School Press, Boston, 1999.

Index